W0043380

Advances in Stereotactic and Functional Neurosurgery 8

Proceedings of the 8ᵗʰ Meeting
of the European Society for Stereotactic
and Functional Neurosurgery,
Budapest 1988

Edited by

G. Broggi, J. Burzaco, E. R. Hitchcock,
B. A. Meyerson, Sz. Tóth

Acta Neurochirurgica
Supplementum 46

Springer-Verlag Wien New York

Prof. Dr. Giovanni Broggi
Istituto Neurologico "C. Besta", Milano, Italy

Dr. Juan Burzaco
Servicio de Neurocirugía, Fundación Iménez Díaz, Madrid, Spain

Professor Edward R. Hitchcock
Department of Neurosurgery, University of Birmingham, U.K.

Prof. Dr. Björn A. Meyerson
Department of Neurosurgery, Karolinska Sjukhuset, Stockholm, Sweden

Dr. Szabolcs Tóth
Department of Neurosurgery, Medical University of Debrecen, Debrecen, Hungary

With 58 Figures

Product Liability: The publisher can give no guarantee for information about drug dosage and application thereof contained in this book. In every individual case the respective user must check its accuracy by consulting other pharmaceutical literature.

This work is subject to copyright

All rights are reserved, whether the whole or part of the material is concerned, specifically those of translation, reprinting, re-use of illustrations, broadcasting, reproduction by photocopying machine or similar means, and storage in data banks.

© 1989 by Springer-Verlag/Wien
Softcover reprint of the hardcover 1st edition 1989

Library of Congress Cataloging-in-Publication Data. European Society for Stereotactic and Functional Neurosurgery. Meeting (8th: 1988: Budapest, Hungary) Advances in stereotactic and functional neurosurgery 8: proceedings of the 8th Meeting of the European Society for Stereotactic and Functional Neurosurgery, Budapest 1988/edited by G. Broggi ... [et al.]. p. cm.—(Acta neurochirurgica. Supplementum. ISSN 0065-1419; 46) ISBN-13: 978-3-7091-9031-9 1. Nervous system—Surgery—Congresses. 2. Stereoencephalotomy—Congresses. I. Broggi, G. (Giovanni) II. Title. III. Series. [DNLM: 1. Brain Neoplasms—surgery—congresses. 2. Epilepsy—surgery—congresses. 3. Movement Disorders—surgery—congresses. 4. Neurosurgery—congresses. 5. Pain—surgery—congresses. 6. Stereotaxic Technics—congresses. W1 AC8661 no. 46/WL 368 E89a 1988] RD593.E97 1988. 617.4'8-dc20. 89-11403

ISBN-13: 978-3-7091-9031-9   e-ISBN-13: 978-3-7091-9029-6
DOI: 10.1007/978-3-7091-9029-6

Preface

The last ten years has witnessed a resurgence of interest in stereotactic surgery although this has been mainly in the field of the comparatively simple stereotactic biopsy of intracranial tumours. There is also evidence of a returning interest in functional neurosurgery other than pain which has always sustained high levels of endeavour.

The present work comprises selected papers from a much larger group of interesting and important communications to the European Society for Stereotactic and Functional Neurosurgery. They represent modern views on a wide variety of stereotactic surgical topics from internationally acclaimed experts in this field. The neurosurgeon who has little or no acquaintance with this fruitful sub-specialty will be surprised to find very broad applications of the technique which is gradually replacing many conventional neurosurgical procedures. This is particularly evident in the papers on tumours but there is also a section on the treatment of vascular disease which marks an extension of neurosurgical practice. The Society has always regarded technical advances as important and some of the most recent developments appear in this book. Finally, an exciting new development of neural transplantation marks the beginning of what may be an important part of neurological surgery in the future.

Stereotactic surgery plays an ever increasing part in the treatment of epilepsy. Jerome Engel Jr., and his colleagues present an over-view of investigations and indications for the management of this disorder and a comprehensive outline of the research protocol from one of the most important centres for the study of epilepsy.

Edward Hitchcock
President

Contents

Listed in Current Contents

Tumours

Vascular Diseases

Technical Progress

Epilepsy

Acta Neurochirurgica, Suppl. 46, 3–8 (1989)
© by Springer-Verlag 1989

Surgical Treatment of Epilepsy: Opportunities for Research Into Basic Mechanisms of Human Brain Function

J. Engel, Jr.[1, 3, 4], **Th. L. Babb**[1, 4], and **P. H. Crandall**[2, 4]

Departments of [1] Neurology, [2] Surgery, [3] Anatomy, and the [4] Brain Research Institute, UCLA School of Medicine, Los Angeles, California, U.S.A.

Summary

Numerous technological developments in neurology have increased the ability to localize structural and functional abnormalities within the human brain. Such techniques have contributed to a renewed interest in resective surgical treatment for medically refractory partial seizures. Enhanced capacity to carry out detailed *in vivo* and *in vitro* measurements of neuronal activity in patients, during the course of presurgical evaluation and following surgical resection, now offers unprecedented opportunities for invasive research into normal and abnormal human cerebral function. Electrophysiological, microanatomical, biochemical and behavioral studies can be carried out without presenting undue risk or discomfort to the patient. Such research in a clinical setting presents difficulties in experimental design for the basic neuroscientist. Problems are reduced in clinical programs where diagnostic and surgical procedures are carried out in a standardized fashion according to specific protocols. The UCLA clinical protocol for anterior temporal lobectomy, based on presurgical evaluation with stereotactically implanted depth electrodes, is particularly amenable to the integration of basic research projects. This protocol and related ongoing research projects are described.

Keywords: Epilepsy surgery; basic research; mechanisms.

Introduction

Surgical resection of epileptogenic brain tissue has been a treatment for recurrent epileptic seizures for over 100 years[24, 40]. Application of this technique was initially limited by the need to correlate a visually identified structural lesion with clinical signs and symptoms. Patient selection improved with the advent of diagnostic radiology, and functional localization of epileptogenic brain tissue became possible following the introduction of electroencephalography[9, 31]. Nevertheless, resective surgical treatment has continued to be available to relatively few patients with medically intractable epilepsy and, by 1986, there were only approximately 50 active epilepsy surgery facilities worldwide, the vast majority of which operated on an average of 10 to 20 patients per year[13]. Technological development in the past decade, however, has included an explosion of diagnostic tools in neurology capable of precisely localizing cerebral disturbances responsible for chronic recurrent partial epileptic seizures.

A recently renewed interest in epilepsy surgery, due in part to these improved techniques for cerebral localization, is backed by a century of data demonstrating that resective surgery can be a safe and effective treatment for partial epilepsy and recognition that this is a greatly underutilized therapeutic modality. An increasing role for resective surgery not only benefits many epileptic patients who otherwise would continue to experience recurrent seizures, but also presents unprecedented opportunities for invasive research into basic mechanisms of normal and abnormal human brain function. *In vivo* studies previously possible only in experimental animals can now be carried out in the course of presurgical evaluation and *in vitro* studies are accomplished with surgically removed tissue, without increased risk or discomfort to the patient[20, 39].

There remain many approaches to the surgical treatment of epilepsy[14, 30], some of which provide better opportunities for the basic neuroscientist than others. Research into fundamental mechanisms is most easily accomplished in epilepsy surgery programs that utilize the scientific method when collecting clinical data. This requires the use of well defined protocols and standardized procedures. The UCLA approach to one type of resective surgical treatment of epilepsy, *anterior temporal lobectomy*, is particularly well suited for basic research projects. The aspects of this approach which lend themselves to valid experimental design, and a brief description of ongoing investigation into fundamental mechanisms are here described.

Clinical Protocol

The epilepsy surgery protocol at UCLA is divided into four phases. Phase one is an inpatient evaluation utilizing EEG telemetry and video monitoring, as well as other extracranial techniques for localization of the epileptogenic region. Phase two is a second inpatient evaluation with intracranial electrodes, Phase three is resective surgery, and Phase four is follow-up. Patients who go from Phase one directly to anterior temporal lobectomy, hemispherectomy, corpus callosum section, or subdural grid evaluation for focal cortical resection will not be considered further here. Although these patients are also subjects for research projects, the bulk of investigations into fundamental mechanisms involves patients with complex partial seizures of suspected limbic origin who proceed to Phase two with stereotactically implanted depth electrodes.

Admission to the Protocol

Patients can be accepted into the epilepsy surgery protocol if there is no medical contraindication to surgery and if their seizures are demonstrated to be of partial onset, have not responded to adequate serum levels of the standard first-line antiepileptic drugs, and significantly interfere with their ability to lead a normal life[1, 19]. Patients are considered potential candidates for anterior temporal lobectomy when they have complex partial seizures of presumed limbic origin, and do not have chronic psychosis or severe mental retardation. In practice, therefore, most of these patients are normal individuals apart from their epileptic seizures and, therefore, acceptable subjects for research into aspects of normal, as well as abnormal, brain function.

Phase One

Patients are usually admitted to the hospital for one to two weeks of EEG telemetry and video monitoring using scalp and sphenoidal electrodes. A complete neuropsychological evaluation is also carried out[19, 32]. If ictal events consistently arise from one sphenoidal electrode, the patient then undergoes thiopental activation, intracarotid sodium amytal (ISA) testing, and positron emission tomography (PET)[19]. These patients can go directly to anterior temporal lobectomy if a focal functional deficit of the temporal lobe initiating habitual seizures is demonstrated by localized hypometabolism on PET, and confirmed by other functional findings such as localizing psychological deficits, focal attenuation of fast activity with thiopental, or memory disturbances with contralateral carotid injection of amytal. Nonspecific structural abnormalities such as dilated ventricles, other evidence of atrophy, or calcification, seen on X-ray computed tomography (XCT) and/or magnetic resonance imaging (MRI), may also be used to confirm ictal EEG findings. Patients who do not meet these criteria for anterior temporal lobectomy, but whose Phase one data continue to suggest a single origin of their habitual seizures, become candidates for Phase two evaluation. Those with complex partial seizures of suspected limbic origin are recommended for depth electrode implanation, while those whose seizures are more likely to be of neocortical origin are recommended for placement of subdural grids[28].

Phase Two

For patients with complex partial seizures of suspected limbic origin, multicontact depth electrodes are implanted bilaterally and symmetrically in a standardized fashion. Targets always include amygdala and anterior hippocampal pes unless there is some contraindication to these placements, as well as four to six other targets on each side chosen from middle and posterior hippocampal pes, anterior middle and posterior hippocampal gyrus, and presubiculum. Bilaterally symmetrical multicontact electrodes are also placed extratemporally when the Phase one evaluation suggests that complex partial seizures could possibly originate from a specific extratemporal limbic site, or propagate into the limbic system from a suspected neocortical epileptogenic region. Bilaterally symmetrical standardized recordings have facilitated interpretation of ictal and interictal depth electrode recordings for clinical purposes[19]. This approach also greatly simplifies experimental designs for basic investigations since patient data can be directly compared, and interhemispheric or interpatient differences can be easily determined. For purely research purposes, many of the depth electrodes placed in mesial temporal structures also contain sheaths of nine fine-wire microelectrodes capable of recording single and multiple unit activity[6]. These microelectrodes have been used routinely at UCLA for 17 years and there is no evidence they present any additional risk to the patient.

The procedure used for implanting depth electrodes at UCLA now makes use of MRI, XCT, and digital subtraction angiography (DSA) performed stereotactically. For most patients, PET data is also obtained using the same stereotactic frame so that metabolic functional images can be correlated precisely with cerebral structures displayed by MRI. Regions of interest can be automatically drawn on the PET scans by tracing anatomical structures on the MRI scans. Non-magnetic metal electrodes are used so post-implantation MRI can confirm the accuracy of placement and the precise recording points for both macro- and microelectrodes can be determined for research purposes.

Following depth electrode implantation, continuous EEG telemetry with video monitoring is carried out for several weeks to capture habitual ictal events. These recordings make use of montages that include symmetrical depth electrode derivations as well as scalp or skull electrodes to monitor surface EEG activity. Thiopental activation and ISA studies are also performed using depth electrode recordings.

Patients are recommended for a standardized anterior temporal lobectomy when Phase two evaluation indicates, with a high degree of confidence, that their habitual seizures begin within the cerebral tissue to be resected. Focal ictal onsets from mesial temporal structures are considered to be reliable localizing findings[19], while regional onsets and occasional contralateral onsets often require the presence of additional confirmatory evidence of primary involvement of one anterior temporal lobe. If Phase two evaluation suggests an extratemporal epileptogenic region that was not adequately defined by depth electrode recordings, and a reasonable localizing hypothesis can be formulated, a second Phase two with subdural grid electrodes might be recommended. No surgery or further evaluation is considered when seizures originate with equal frequency from either hemisphere, or when diffuse or bilaterally synchronous ictal onsets are encountered.

Phase Three

When the patient meets the criteria for a standardized anterior temporal lobectomy, this is performed *en bloc* under generalized anesthesia without intraoperative electrocorticography (ECoG). ECoG and functional mapping are rarely performed when there is question concerning the posterior extent of the epileptogenic region since these patients now usually undergo extraoperative evaluation with subdural grid recording. The technique for, and dimensions of, the *en bloc* resection differ slightly for left and right anterior temporal lobes[11, 12, 22] but are relatively consistent from one pa-

tient to another. This permits the entire specimen to be easily oriented by the pathologist when choosing tissue for *in vitro* investigation. The recording sites of depth electrodes can be identified, and correlation with stereotactically acquired structural and functional imaging data can be made.

Phase Four

Patients who undergo anterior temporal lobectomy receive comprehensive follow-up evaluations. Postoperative MRI scans are carried out to determine completeness of resection. Every attempt is made to bring all patients back to UCLA at one, two, five and ten years after surgery for complete neurologic evaluation, EEG, and psychometric testing. Where possible, patients return at three months and six months postoperatively, and then at yearly intervals, for routine neurological evaluation. In addition, an independent postoperative follow-up team consisting of a nurse, an occupational therapist and a social worker maintain contact with the patient to evaluate their psychosocial progress following surgery, and to elicit medical concerns or complaints that patients might be unwilling to tell their doctor. Psychiatric follow-up and intervention is available when indicated. The postoperative course with respect to seizures helps determine whether the resected tissue was responsible for, or involved in, the habitual ictal events. Other information is used to determine whether removal of normal tissue within the resected specimen might have resulted in neurological deficit, and whether removal of epileptogenic activity might have had such beneficial effects on interictal behavior as improved memory and increased IQ[36].

Research Protocol

Experimental designs must take into account uncertainties in distinguishing between normal and abnormal brain. It can be argued that no data obtained from the brains of individuals with severe partial epilepsy should be considered truly normal[16]. Consequently, interpretation of research into fundamental mechanisms of normal brain function must include consideration of possible contamination by epileptiform disturbances, while interpretation of research into fundamental mechanisms of epilepsy is limited by the fact that control data can be extremely difficult, or perhaps impossible, to obtain. A strategy for overcoming the difficulties imposed by this clinical reality has been to perform correlative investigations involving as many different functional and structural measurements from the same tissue as possible. Consequently, *in vivo* electrophysiological, metabolic, and behavioral investigations carried out during the course of presurgical evaluation are correlated with each other, and also with *in vitro* microanatomical, biochemical, and electrophysiological investigations carried out postoperatively.

For all investigations, the epileptogenic region is defined as the area demonstrated electrophysiologically to give rise to most or all of the patient's habitual ictal events. This is confirmed in most cases by the presence of a structural lesion, usually hippocampal sclerosis[3], in the resected specimen and a postoperative marked reduction or disappearance of spontaneous epileptic seizures. *In vivo* data obtained from the contralateral hemisphere have been used to study "normal" phenomena, although it is realized that functional, if not structural, disturbances related to the epileptic condition certainly exist in at least some patients[18]. Furthermore, outcome with respect to epileptic seizures does not definitively indicate that the primary epileptogenic region was included within the resected brain tissue: seizures may cease because an important pathway was removed or a critical mass of tissue was reduced, while some seizures may continue after removal of the primary region because areas of residual brain tissue retain some degree of epileptogenicity[15].

Mechanisms of Normal Human Brain Function

Microelectrode recordings from the hippocampus contralateral to the primary epileptogenic region have revealed unit firing patterns with burst characteristics and periodicities that resemble, with some exceptions, the behavior of hippocampal neurons in lower animals[5, 25]. Visual and other sensory responses of some neurons in mesial temporal lobe structures have been demonstrated and receptive fields of lateral geniculate neurons have been characterized by recording from optic radiation fibers passing through posterior temporal lobe[42].

Intrahemispheric connections between mesial temporal limbic structures have been defined electrophysiologically, and paired pulse facilitation was found to exist in the human hippocampus as it does in lower animals[43]. Evidence suggests, however, that the hippocampal commissure is not a functional pathway in the human[44], explaining why epileptic seizures beginning in one hippocampus characteristically persist for long periods of time before spreading to the contralateral hemisphere[26, 27].

Characteristics of continuous or ISA-induced memory deficits have been correlated with patterns of cell loss in hippocampus, as well as the extent of focal hypometabolism on PET scans. These findings suggest regional specificity within mesial temporal structures for different types of memory processing[33–35].

Studies of relationships between amygdala unit firing in the nonepileptogenic temporal lobe and cardiorespiratory function have allowed direct comparison with animal data on amygdala control of heart rate and respiration[23]. Similar studies of the epileptogenic temporal lobe, combined with information from EEG and cardiorespiratory monitoring during interictal spikes and ictal events, as well as during amygdala stimulation, may provide insights into the mechanisms of sudden unexplained death in some patients with epilepsy[23].

Mechanisms of Epileptogenic Human Brain Dysfunction

Routine interictal depth electrode recordings have demonstrated that the epileptogenic abnormality in patients with medically refractory complex partial seizures is not focal. Rather, interictal spikes commonly arise from an extensive area of brain ipsilateral to the site of ictal onset, and independent contralateral spikes are also commonly encountered[16]. PET scans also indicate that the functional disturbance is widespread[16]. A variety of patterns of depth electrode recorded ictal onsets can be seen and these often consist of high amplitude spike-and-wave discharges, rather than low voltage fast activity. The former suggest that hypersynchrony may be a mechanism of spontaneous seizure generation[16, 21].

Careful microanatomical studies of resected temporal lobes have allowed the degree and extent of hippocampal cell loss to be quantitatively determined for correlation with electrophysiological, biochemical and behavioral data[4]. Even 30 percent reduction in the density of hippocampal cells might account for epileptogenicity in some patients[17]. Morphological disturbances of pyramidal cells in epileptogenic hippocampus have also been described[3, 37], indicating a loss in dendritic spines and a reduction in dendritic domain similar to neuronal changes observed in the experimental alumina focus[41].

Correlations between ictal depth electrode recordings and postoperative microanatomical analyses indicate that seizures originate in regions of hippocampus where cell loss has taken place, and not in the more normal appearing surrounding tissue[7]. *In vivo* stimulation studies have demonstrated that this sclerotic tissue has a higher threshold for electrically induced after-discharge[10], and that paired pulse suppression, rather than facilitation, is commonly encountered[21, 43]. These findings have been taken as evidence for enhanced, rather than diminished, inhibitory mechanisms within the interictal epileptogenic region[21], a conclusion supported by immunohistochemical studies which have failed to demonstrate a reduction of GABAergic terminals or interneurons within the cell sparse region where epileptic seizures begin[2].

In vivo unit recordings from epileptogenic hippocampus have revealed characteristic burst firing followed by suppression, temporally associated with the locally recorded EEG spike-and-wave[6]. This observation suggests that mechanisms of human hippocampal epilepsy are similar to those producing paroxysmal depolarization shifts and afterhyperpolarizations in experimental animal models[28]. However, the percentage of units participating in this abnormal activity in the human epileptogenic hippocampus is much lower than that described for the acute penicillin focus of cat neocortex. Unit recordings during ictal onset in the human hippocampus demonstrate that firing of many cells may actually decrease in a manner that suggests hypersynchronization, rather than hyperexcitation, as a common mechanism of seizure generation[8]. Further studies using *in vitro* electrophysiological techniques[35] are now underway at UCLA in an attempt to identify and define reorganization of local circuits in sclerotic hippocampus which might account for the patterns of epileptogenicity observed *in vivo*.

Acknowledgements

Research reported here has been supported in part by Grants NS-02808 and NS-15654 from the National Institutes of Health, Contract DE-AC 03-76-SF 000112 from the Department of Energy and the David H. Murdock Foundation for Advanced Brain Studies.

References

1. Andermann F (1987) Identification of candidates for surgical treatment of epilepsy. In: Engel J Jr (ed) Surgical treatment of the epilepsies. Raven Press, New York, pp 51–70
2. Babb TL (1986) GABA-mediated inhibition in the Ammon's Horn and presubiculum in human temporal lobe epilepsy: GAD immunocytochemistry. In: Nisticò G, Morselli PL, Lloyd KG, Fariello RG, Engel Jr J (eds) Neurotransmitters, seizures, and epilepsy III. Raven Press, New York, pp 293–302
3. Babb TL, Brown WJ (1987) Pathological findings in epilepsy. In: Engel Jr J (ed) Surgical treatment of the epilepsies. Raven Press, New York, pp 511–540
4. Babb TL, Brown WJ, Pretorius J, Davenport C, Lieb JP, Crandall PH (1984) Temporal lobe volumetric cell densities in temporal lobe epilepsy. Epilepsia 25: 729–740
5. Babb TL, Carr E, Crandall PH (1973) Analysis of extracellular firing patterns of deep temporal lobe structures in man. Electroenceph Clin Neurophysiol 34: 247–257
6. Babb TL, Crandall PH (1976) Epileptogenesis of human limbic neurons in psychomotor epileptics. Electroenceph Clin Neurophysiol 40: 225–243
7. Babb TL, Lieb JP, Brown WJ, Pretorius J, Crandall PH (1984) Distribution of pyramidal cell density and hyperexcitability in the epileptic human hippocampal formation. Epilepsia 25: 721–728
8. Babb TL, Wilson CL, Isokawa-Akesson M (1987) Firing patterns of human limbic neurons during stereoencephalography (SEEG) and clinical temporal lobe seizures. Electroenceph Clin Neurophysiol 66: 467–482
9. Bailey P, Gibbs FA (1951) Surgical treatment of psychomotor epilepsy. J Am Med Assoc 145: 365–370
10. Cherlow DG, Dymond AM, Crandall PH, Walter RD, Serafetinides EA (1977) Evoked response and after-discharge thresholds to electrical stimulation in temporal lobe epileptics. Arch Neurol 34: 527–531
11. Crandall PH (1987) Cortical resection. In: Engel Jr J (ed) Surgical treatment of the epilepsies. Raven Press, New York, pp 377–404
12. Crandall PH, Walter RD, Rand RW (1963) Clinical applications of studies on stereotactically implanted electrodes in temporal lobe epilepsy. J Neurosurg 20: 827–840
13. Engel Jr J (ed) (1987) Surgical treatment of the epilepsies. Raven Press, New York, pp 727
14. Engel Jr J (1987) Approaches to localization of the epileptogenic lesion. In: Engel Jr J (ed) Surgical treatment of the epilepsies. Raven Press, New York, pp 75–96
15. Engel Jr J (1987) Outcome with respect to epileptic seizures. In: Engel Jr J (ed) Surgical treatment of the epilepsies. Raven Press, New York, pp 553–572
16. Engel Jr J (1987) New concepts of the epileptic focus. In: Weiser HG, Speckmann EG, Engel Jr J (eds) The epileptic focus. John Libbey Eurotext Ltd, London, pp 83–94
17. Engel Jr J, Babb TL, Phelps ME (1987) Contribution of positron emission tomography to understanding mechanisms of epilepsy. In: Engel Jr J, Ojemann GA, Lüders HO, Williamson PD (eds) Fundamental mechanisms of human brain function. Raven Press, New York, pp 209–218
18. Engel Jr J, Crandall PH (1983) Falsely localizing ictal onsets with depth EEG telemetry during anticonvulsant withdrawal. Epilepsia 24: 344–355
19. Engel Jr J, Crandall PH, Rausch R (1983) The partial epilepsies. In: Rosenberg RN, Grossmann RG, Schochet S, Heinz ER, Willis WD (eds) The clinical neurosciences, vol 2. Churchill Livingstone, New York, pp 1349–1380
20. Engel Jr J, Ojemann G, Lüders HO, Williamson PD (eds) (1987) Fundamental mechanisms of human brain function. Raven Press, New York, pp 288
21. Engel Jr J, Wilson CL (1986) Evidence for enhanced synaptic inhibition in epilepsy. In: Nisticò G, Moselli PL, Lloyd KG, Fariello RG, Engel Jr J (eds) Neurotransmitters, seizures and epilepsy, III. Raven Press, New York, pp 1–13
22. Falconer MA (1967) Surgical treatment of temporal lobe epilepsy. NZ Med J 66: 539–542
23. Frysinger RC, Harper RM, Hackel RJ (1987) State-dependent cardiac and respiratory changes associated with complex partial epilepsy. In: Engel Jr J, Ojemann GA, Lüders HO, Williamson PD (eds) Fundamental mechanisms of human brain function. Raven Press, New York, pp 219–226

24. Horsley V (1886) Brain-surgery. Br Med J 2: 670–675
25. Isokawa-Akesson M, Babb TL, Wilson CL (1987) Physiology of hippocampal neurons in humans and lower mammals. In: Engel Jr J, Ojemann GA, Lüders HO, Williamson PD (eds) Fundamental mechanisms of human brain function. Raven Press, New York, pp 15–25
26. Lieb JP, Babb TL, Engel Jr J, Darcey TM (1987) Propagation pathways of interhemispheric seizure discharges compared in human and animal hippocampal epilepsy. In: Engel Jr J, Ojemann GA, Lüders HO, Williamson PD (eds) Fundamental mechanisms of human brain function. Raven Press, New York, pp 165–170
27. Lieb JP, Engel Jr J, Babb TL (1986) Interhemispheric propagation time of human hippocampal seizures. I. Relationship to surgical outcome. Epilepsia 27: 286–293
28. Lüders H, Lesser RP, Dinner DS, Morris HH, Hahn JF, Friedman L, Skipper G, Wyllie E, Friedman D (1987) Commentary: chronic intracranial recording and stimulation with subdural electrodes. In: Engel Jr J (ed) Surgical treatment of the epilepsies. Raven Press, New York, pp 297–322
29. Matsumoto H, Ajmone-Marsan C (1964) Cortical cellular phenomena in experimental epilepsy: Interictal manifestations. Exp Neurol 9: 286–304
30. Ojemann GA, Engel Jr J (1987) Acute and chronic intracranial recording and stimulation. In: Engel Jr J (ed) Surgical treatment of the epilepsies. Raven Press, New York, pp 263–288
31. Penfield W, Jasper H (1954) Epilepsy and the functional anatomy of the human brain. Little Brown and Company, Boston, pp 896
32. Rausch R (1987) Neurophysiological evaluation. In: Engel Jr J (ed) Surgical treatment of the epilepsies. Raven Press, New York, pp 181–196
33. Rausch R, Babb TL (1987) Evidence for memory specialization within the mesial temporal lobe in man. In: Engel Jr J, Ojemann GA, Lüders HO, Williamson PD (eds) Fundamental mechanisms of human brain function. Raven Press, New York, pp 103–109
34. Rausch R, Babb TL (in press) Extent and pattern of hippocampal damage and verbal memory deficits. Brain
35. Rausch R, Babb TL, Engel Jr J, Crandall PH (in press) Memory following intracarotid sodium amytal injection contralateral to hippocampal damage. Arch Neurol
36. Rausch R, Crandall PH (1982) Psychological status related to surgical control of temporal lobe seizures. Epilepsia 23: 191–202
37. Scheibel ME, Crandall PH, Scheibel AB (1974) The hippocampal-dentate complex in temporal lobe epilepsy. Epilepsia 15: 55–80
38. Schwartzkroin PA (1987) The electrophysiology of human brain slices resected from "epileptic" brain tissue. In: Engel Jr J, Ojemann GA, Lüders HO, Williamson PD (eds) Fundamental mechanisms of human brain function. Raven Press, New York, pp 145–154
39. Sherwin AL (in press) Guide to neurochemical analysis of surgical specimens of human brain. Epilepsy Research
40. Taylor DC (1987) One hundred years of epilepsy surgery: Sir Victor Horsley's contribution. In: Engel Jr J (ed) Surgical treatment of the epilepsies. Raven Press, New York, pp 7–11
41. Westrum LE, White LE Jr, Ward AA Jr (1964) Morphology of the experimental epileptic focus. J Neurosurg 21: 1033–1046
42. Wilson CL (1987) Microelectrode recordings of human temporal lobe neurons during sensory stimulation. In: Engel Jr J, Ojemann GA, Lüders HO, Williamson PD (eds) Fundamental mechanisms of human brain function. Raven Press, New York, pp 1–13
43. Wilson CL, Isokawa-Akesson M, Babb TL, Engel Jr J (1986) Hippocampal field-potential and neuronal response facilitation and depression during paired-pulse stimulation in humans. Neurosci Abst 12: 1330
44. Wilson CL, Isokawa-Akesson M, Babb TL, Engel Jr J, Cahan LD, Crandall PH (1987) A comparative view of local and interhemispheric limbic pathways in humans: an evoked potential analysis. In: Engel Jr J, Ojemann GA, Lüders HO, Williamson PD (eds) Fundamental mechanisms of human brain function. Raven Press, New York, pp 27–38

Correspondence: Jerome Engel, Jr., Reed Neurological Research Center, UCLA School of Medicine, Los Angeles, California, CA 90024, U.S.A.

Acta Neurochirurgica, Suppl. 46, 9–12 (1989)
© by Springer-Verlag 1989

Stereotactic Investigations in Frontal Lobe Epilepsies

C. Munari[1, 2], A. T. Giallonardo[1, 3], P. Brunet[1], D. Broglin[2], and J. Bancaud[1]

[1] INSERM U 97, Paris, France, [2] Hôpital Sainte Anne, Service de Neurochirurgie, Paris, France, [3] Va Clinica Neurologica "La Sapienza", Roma, Italy

Summary

The aim of a Stereo-EEG investigation is to verify and prove that the hypothesis, done on the basis of the preliminary investigations (clinical, EEG, neuroradiological), are correct. This task is particularly hard in frontal lobe epilepsies, because of anatomical and physiopathological reasons. Among 277 consecutive patients, 86 were explored for a probable frontal epilepsy. The stereotactically introduced electrodes,

1) simultaneously record the electrical activity on both, mesial and lateral cortical areas, and,

2) in $^3/_4$ of cases also investigate extra-frontal, mainly temporal, areas.

Two small, non-surgical haematomas were provoked in one patient.

The spatial trajectory of the discharges, evaluated with this methodology, permits of limiting the surgical removal in many cases.

Keywords: Depth electrodes; epilepsy surgery; frontal lobe epilepsy; partial epilepsy; stereo-electro-encephalo-graphy (SEEG); stereotactic surgery.

Introduction

The principles of surgery for partial epilepsies have not changed since the pioneer interventions of Horsley in the 1800s[8]. The neurosurgeon must identify where the seizures start, and define the spatial and functional relationships between the anatomical lesion(s) and the epileptogenic area(s). If the epileptogenic area is "unique and stable" and its removal possible without producing new deficits, then surgery can be advised.

Different investigational approaches are used in major centers preoperatively[10, 29].

In our Department, stereo-EEG investigations with several multilead depth electrodes[13], are considered as both, the end of all preliminary investigations (clinical, EEG, neuroradiological), and the start point for programming surgery[2, 4, 5, 6, 14, 24, 25]. Intracerebral electrode implantation is very dependant on individual anatomo-electro-clinical characteristics.

Many papers are devoted to the pre-surgical and surgical management of the temporal epilepsies, but papers on pre-operative investigations for frontal lobe epilepsies are relatively rare, despite the fact that it is the second most commonly described epilepsy in neurosurgical series[18, 19, 20, 25]. Several explanations were given for this phenomenon, as "the protean manifestations of frontal lobe seizures" rending "meaningful interpretation difficult"[30].

We discuss, on the basis of our experience, if and how the stereo-EEG may be useful for investigating frontal lobe epilepsies.

Material and Methods

Patients: During the period 1974–1985, 300 stereo-EEG investigations were carried out in 277 patients (aged 5 to 50 years; mean age: 24 years) with severe, drug resistant, partial epilepsy. The clinical history was of 2 to 24 years duration before admission.

Surgical procedures: The methodology employed has been described previously[2, 4, 5, 6, 13, 14, 25]. We must stress that the cerebral areas are:

● identified using the stereotactic Atlases of Talairach *et al.*[26, 27, 28] and Szikla *et al.*[23];

● percutaneously reached after stereoscopic angiography and ventriculography and electrocoagulation of the dura.

Anatomical limits: The posterior boundary of the frontal lobe is generally fixed at the central sulcus. However, since the partial motor epilepsies appear to have peculiar characteristics, we arbitrarily fix, the anatomical limits of the frontal lobe, at the pre-central gyrus. We include in the frontal lobe, the suprasylvian precentral opercular region anteriorly to the vertical of the Anterior Commissure (V.C.A.[23, 28]). This option was taken because of the anatomo-functional relationships between these areas and the temporal and peri-insular structures.

Results

Frontal explored areas: Among the 2,326 electrodes (E) implanted for 300 stereo-EEG investigations, 776 (33%) investigated frontal areas. At least one frontal E was implanted in about 90% of patients. Table 1 shows the frontal explored areas in 86 patients with more than three frontal implanted electrodes. In most cases the same electrode simultaneously records the electrical activity from the medial and lateral cortex, avoiding the ventricular frontal horn. In 39 patients the frontal pole was investigated with an antero-posterior ortogonal electrode, at 10 to 24 mm from the individual midline, stopping in front of the projection of the V.C.A.

Mesial structures such as supplementary Motor Area or Anterior Cingulate Gyrus are directly reached by a lateral orthogonal approach: the deepest lead is at 1 to 5 mm from the individual midline according to the midline vessels trajectory.

Bilateral frontal lobes were explored in 48% of patients.

Extra-frontal explored areas (Table 2). In 70% of the 86 patients, we also explored extra-frontal areas, mainly anterior temporal. The electrodes, with deepest contacts in the Amygdala and Ammon's horn record simultaneously, with their superficial contacts, the 2nd temporal convolution[13]. Similarly the parietal cingulate gyrus (and the medial parietal cortex) are investigated by the same multilead electrodes exploring the lateral parietal cortex.

Complication: In this series, one patient presented one small haematoma in both frontal lobes.

Discussion

The choice between "invasive" or "non-invasive" investigations, before deciding to operate, is matter of great discussion[11, 29]. In some of the most experienced centres in the world (*e.g.* Montreal Neurological Institute), the use of intracerebral recordings pre-operatively is reserved for a small percentage of patients, where there is an ambiguity of lateralization, secondary generalized or multifocal patterns, or predominant discharges distant from, or controlateral to, a structural lesion[12].

In our experience, stereotactically implanted electrodes must give enough information to decide not only the indications for operation, but also the anatomical limits of the excision[7].

There are several criticisms of intracerebral recordings. Firstly the technique is "invasive". In fact, careful use of the stereotactic stereoscopic angiography[15, 23] avoids the risk of severe operative haemorrhage[13], and has a very low complication rate similar to those due to subdural strip electrodes[31]. The second criticism concerns the relatively high number of implanted electrodes. The correct study of the spatial and temporal trajectory of the ictal discharges in the brain, requires simultaneous records from both the medial and lateral cortex, which is even more important in frontal lobe epilepsies, because of the "complexity" of the clinical manifestations[1, 3, 10, 16, 19, 30]. Moreover, since the motor symptoms are variable and may occur at different stages of the seizures, it is important to verify the possible relationships between onset of discharge, spread and clinical course. Another major problem is identification of the side where discharges start: this explains why both frontal lobes were explored in 48% of patients, while in temporal epilepsies bilateral investigations are 10%[13].

It is relatively difficult to explore the medial postero-inferior part of the orbital cortex with our orthogonal approach, whereas the lateral approach allows a correct investigation of the suprasylvian frontal opercular region, provided stereoscopic angiography is done.

Table 1. *Frontal Lobe Epilepsies (86 Patients)*

	Stereotactically explored structures	Electrodes
F	internal frontal cortex	281
R	2nd frontal convolution	258
O	3rd frontal convolution	209
N	1st frontal convolution	24
T	orbital cortex	164
A	cingulate gyrus	133
L	frontal pole (A.P.)	39
	bilateral stereo-eeg: 42 (48%)	
	right/left ratio: 2	

Table 2. *Frontal Lobe Epilepsies (86 Patients)*

Extrafrontal explored structures ($^{61}/_{86}$ Patients)	Electrodes
Temporal	
Pole	7
Amygdala	18
Ammon's horn	24
T 1	14
T 2	42
Parietal	
Cingulate gyrus	7
Medial cortex	20
Lateral cortex	27

Even using a relatively high number of intracerebral electrodes, the possibility of correctly identify a small "epileptogenic" region is less frequent than in temporal lobe epilepsies, probably because of the anatomofunctional organization of the frontal areas[9, 17]. In several cases we can limit the surgical removal to the external, or internal frontal cortex, thus avoiding a large and unnecessary frontal lobectomy. Therefore to investigate a frontal lobe with only 2 or 3 electrodes is insufficient for correct definition of the epileptogenic area[21], even if so few electrodes can show that the seizures start in a frontal lobe.

The simultaneous implantation of electrodes in both, frontal and temporal areas allowed us to understand better the complex relationships between the anterior-temporal and basal frontal areas[16, 22]. Subjective manifestations, such as epigastric "aura", generally considered as a synonimous of an initial temporal lobe discharge, may actually be the first sign of a frontal seizure, with somatomotor lateralized manifestations, or of a temporal seizure with oro-alimentary automatisms (Munari *et al.* unpublished data, presented at the Esclimont Meeting in November 1987).

Only the simultaneous recording of both frontal and temporal structures permits a correct differential diagnosis in such cases.

References

1. Ajmone Marsan C, Ralston B (1957) The epileptic seizure. Its functional morphology and diagnostic significance. Charles C Thomas, Springfield, Ill
2. Bancaud J (1959) Apport de l'exploration fonctionnelle par voie stéréotaxique à la chirurgie de l'épilepsie. Neurochirurgie 5: 55–112
3. Bancaud J (1967) Origine focale multiple de certaines epilepsies corticales. Rev Neurol 117: 222–243
4. Bancaud J, Talairach J (1970) L'électroencéphalographie de profondeur (SEEG dans l'épilepsie). In: Modern problems of pharmaco-psychiatry epilepsy, vol 4. Karger, Basel New York, pp 29–41
5. Bancaud J, Talairach J, Bonis A *et al* (1965) La stéréo-électro-encéphalographie dans l'épilepsie. Masson et Cie, Paris
6. Bancaud J, Talairach J, Geier S, Scarabin JM (1973) EEG et SEEG dans les tumeurs cérébrales et l'épilepsie. Edifor, Paris
7. Centre Saint Paul Broca, Paris (1987) Appendix II: Presurgical evaluation protocols. In: Engel Jr J (ed) Surgical treatment of epilepsies. Raven Press, New York, pp 671–672
8. Crandall PH (1987) Cortical resections. In: Engel Jr J (ed) Surgical treatment of the epilepsies. Raven Press, New York, pp 377–404
9. Creutzfeldt OD, Wieser HG (1987) Physiology of the frontal cortex. In: Wieser HG, Elger CE (eds) Methods of presurgical evaluation of epileptic patients. Springer, Berlin Heidelberg New York, pp 23–27
10. Delgado-Escueta AV, Swartz BE, Maldonado HM, Walsh GO, Rand RW (1987) Complex partial seizures of frontal lobe origin. In: Wieser HG, Elger CE (eds) Methods of presurgical evaluation of epileptic patients. Springer, Berlin Heidelberg New York, pp 267–299
11. Engel Jr J (1987) Surgical treatment of the epilepsies. Raven Press, New York, 727 pp
12. Montreal Neurological Hospital and Institute (1987) Appendix II: Presurgical evaluation protocols. In: Engel Jr J (ed) Surgical treatment of epilepsies. Raven Press, New York, pp 682–686
13. Munari C (1987) Depth electrode implantation at Hospital Sainte Anne, Paris. In: Engel Jr J (ed) Surgical treatment of the epilepsies. Raven Press, New York, pp 583–588
14. Munari C, Bancaud J (1985) The role of stereo-electro-encephalography (SEEG) in the evaluation of partial epileptic seizures. In: Porter RJ, Morselli PL (eds) The epilepsies. Butterworths, Sevenoaks, pp 267–306
15. Munari C, Giallonardo AT, Musolino A, Brunet P, Chodkiewicz JP, Bancaud J, Talairach J (1987) Specific neuroradiological examinations necessary for stereotactic procedures. In: Wieser HG, Elger CE (eds) Methods of presurgical evaluation of epileptic patients. Springer, Berlin Heidelberg New York, pp 141–145
16. Munari C, Stoffels C, Bossi L, Bonis A, Talairach J, Bancaud J (1981) Automatic activities during frontal and temporal lobe seizures: are they the same? In: Dam M, Gram L, Penry JK (eds) Advances in epileptology, XIIth Epilepsy International Symposium. Raven Press, New York, pp 287–291
17. Pandya DN, Yeterian EH (1987) Hodology of limbic and related structures: cortical and commissural connections. In: Wieser HG, Elger CE (eds) Methods of presurgical evaluation of epileptic patients. Springer, Berlin Heidelberg New York, pp 3–14
18. Penfield W, Jasper H (1954) Epilepsy and the functional anatomy of the human brain. Little, Brown & Co, Boston, Mass
19. Rasmussen T (1975) Surgery of frontal lobe epilepsy. In: Purpura DP, Penry JK, Walter RD (eds) Advances in neurology, vol 8. Raven Press, New York, pp 197–204
20. Rasmussen T (1983) Surgical treatment of complex partial seizures: results, lessons and problems. Epilepsia 24, S 1: S 65–S 76
21. Spencer DD (1987) Depth electrode implantation at Yale University. In: Engel Jr J (ed) Surgical treatment of epilepsy. Raven Press, New York, pp 603–607
22. Stoffels C, Munari C, Brunie-Lozano E, Bonis A, Bancaud J, Talairach J (1980) Manifestations automatiques dans les crises épileptiques partielles complexes d'origine frontale. Prog Epileptol 29/30: 111–113
23. Szikla G, Bouvier C, Hori T, Petrov V (1977) Angiography of the human brain cortex. Atlas of vascular patterns and stereotactic cortical localization. Springer, Berlin Heidelberg New York
24. Talairach J, Bancaud J (1973) Stereotactic approach to epilepsy. Methodology of anatomofunctional stereotaxic investigations. Progr Neurol Surg 5: 294–354
25. Talairach J, Bancaud J, Szikla G *et al* (1974) Approche nouvelle de la neurochirurgie de l'épilepsie. Méthodologie stéréotaxique et résultats thérapeutiques. Neurochirurgie [Suppl] 1: 240 pp
26. Talairach J, David M, Tournoux P (1958) L'exploration chirurgicale stéréotaxique du lobe temporal dans l'épilepsie temporale. Masson et Cie, Paris

27. Talairach J, David M, Tournoux P, Corredor H, Kvasina T (1957) Atlas d'anatomie stéréotaxique des noyaux gris centraux. Masson et Cie, Paris

28. Talairach J, Szikla G, Tournoux P *et al* (1967) Atlas d'anatomie stéréotaxique du télencéphale. Masson et Cie, Paris

29. Wieser HG, Elger CE (1987) Presurgical evaluation of epileptics. Basis, techniques, implications. Springer, Berlin Heidelberg New York, 389 pp

30. Williamson PD, Wieser HG, Delgado-Escueta AV (1987) Clinical characteristics of partial seizures. In: Engel Jr J (ed) Surgical treatment of epilepsies. Raven Press, New York, pp 101–120

31. Wyler AR, Ojemann GA, Littich E, Ward AA (1984) Subdural strip electrodes for localizing epileptogenic foci. J Neurosurg 60: 1195–1200

Correspondence: C. Munari, INSERM U 97, 2 ter rue d'Alésia, F-75014 Paris, France.

Acta Neurochirurgica, Suppl. 46, 13–16 (1989)
© by Springer-Verlag 1989

Propagation of Focal Epilepsies and the Timing and Targeting of Surgical Interventions. Illustrative Case Reports

P. Halász, Sz. Tóth, J. Vajda, Gy. Rásonyi, and **O. Eisler**

National Institute for Nervous and Mental Diseases, Budapest, Hungary

Summary

The mirror focus conception assumes propagation of epileptic excitation towards a homotopic area of the contralateral hemisphere inducing secondary focal discharges. Later, independent discharges can be initiated by this new area of epileptogenesis but initiation of seizures of the same or altered symptomatology—and not only interictal discharges—from a mirror focus has yet not been proved. Three cases will be presented where dynamism of the development of temporal lobe mirror focus have been followed. All cases give rise to several questions concerning selection and timing of surgical approach.

Keywords: Mirror focus; temporal lobe surgery; secondary synchronization; callosotomy.

Mirror focus and secondary synchronization are two important mechanisms of action in propagation of epileptic excitation.

The concept of a mirror focus assumes propagation of epileptic excitation towards a homotopic area of the contralateral hemisphere inducing secondary focal discharges. Later independent discharges can be initiated by this new area of epileptogenesis but initiation of seizures of the same or altered symptomatology—and not only interictal discharges—from a mirror focus are unproven (Morell 1985, Goldensohn 1984).

Three cases are presented where dynamism of the development of temporal lobe mirror focus has have been followed and which raise several questions concerning selection and timing of surgery.

In the *first case* (26 years, female) during 8 years observation the interictal pattern of active spiking was a constant morphological feature over the right temporal lobe. The corresponding partial seizures were highly stereotyped gestural automatisms with short duration. In the course of the 8 years a mirror focus on the left side could be revealed interictally by hexobarbital sleep while the right temporal spiking remained unchanged even during natural NREM and REM sleep (Fig. 1).

However, long-term EEG monitoring (split screen technic) showed a substantial left side participation in the seizures (Fig. 2). Since the symptomatology of the seizures has been unchanged from the beginning it has to be assumed that this left side participation was present in the seizures before the interictal mirror focus could be detected.

In the *second patient* (17 years, female) an interictal temporal mirrorfocus was present after a four years history of partial seizures characterized by rising gastric sensation, compressing feeling in the

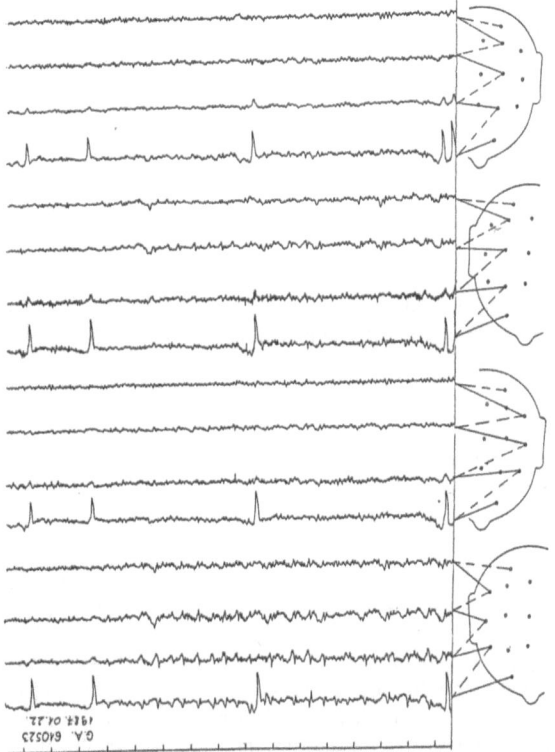

Fig. 1. Patient 1. Trains of interictal sharp waves over the right anterior temporal region in awake state. (Continuous finding during awake, slow wave and REM sleep)

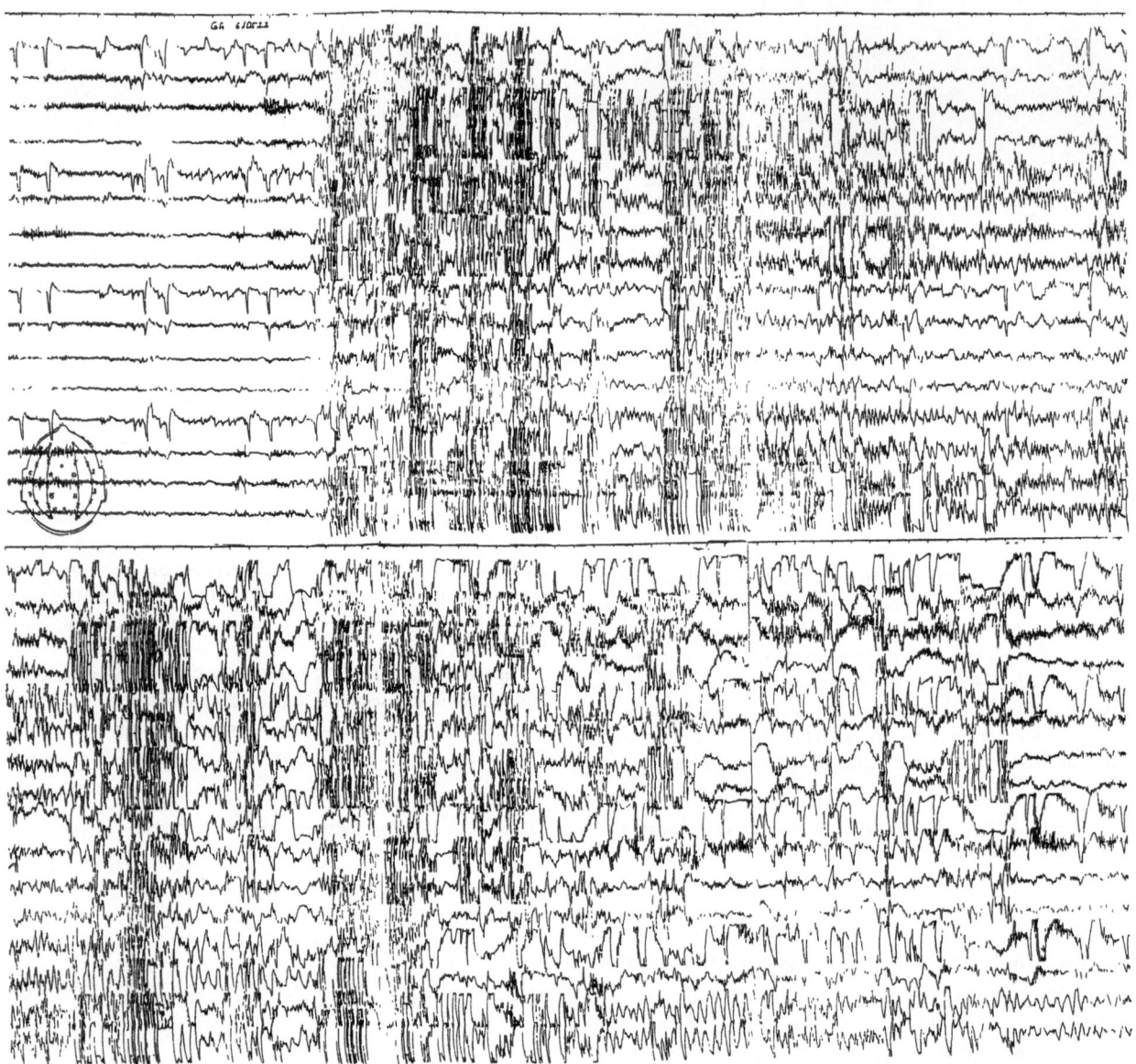

Fig. 2. Patient 1. Ictal discharges during a complex partial seizure characterized by emotional vocalization, gesturing and confusion. The left side preponderance in the ictal pattern is evident

throat, sometimes smells and fear with escape behaviour. During the seizures right anterior temporal ictal discharges regularly appeared. No local pathology was revealed by CT scan.

To treat the pharmacologically intractable seizures right amygdalectomy and partial hippocampectomy was performed giving significant improvement. However, rare seizures remained in the same pattern as before the operation. The interictal right side spiking disappeared but a contralateral spike focus was detected during hexobarbital anaesthesia (Fig. 3). The origin of the seizures could not be localized to the right anterior temporal region since rare seizures still appear today in the same form after the excision of this region. At the same time the role of this region can not be completely denied and it seems to be important in the organization of the seizures otherwise the postoperative improvement can not be explained.

The *third case* (19 years, male) had 3 years recurrent complex

partial seizures consisting of dreamy states with kinaesthetic sensations, verbal automatisms and rare generalized tonic-clonic seizures. During that time a graphomanic attitude and viscosity in his thinking and depressive mood changes with suicidal thoughts developed.

The first interictal EEG showed equivocal left random sharp wave complexes, and ictal discharges seemed to start on the left side, but rapidly spread to the right side. However, in a record taken nine months later bilateral independent sharp wave foci appeared in the temporal region and the ictal pattern in the same record showed bitemporal involvement with different morphological features on each side suggesting right side preponderance. The ictal pattern was followed again by interictal discharges on the left side.

A CT scan (Fig. 4) has revealed a small tumour with a large cyst in the right temporal lobe which was operated upon and proved to

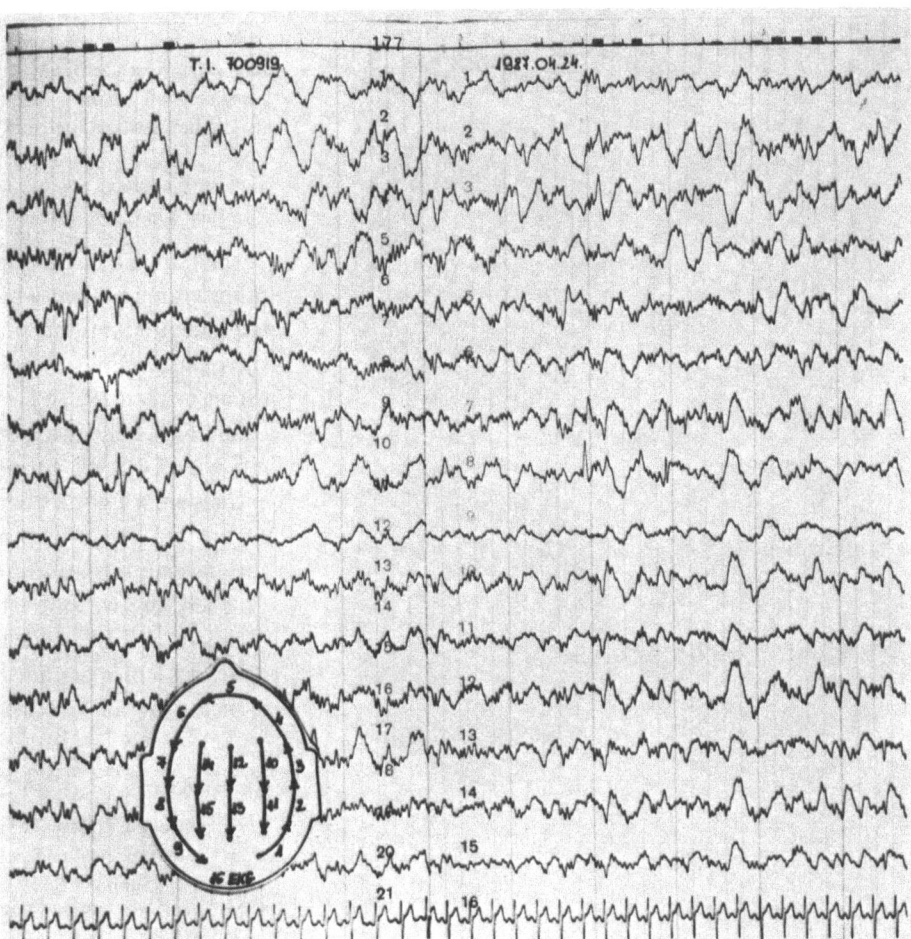

Fig. 3. Patient 2. Left temporal interictal spike focus in hexobarbital sleep 18 months after right temporal partial lobectomy

be a pylocystic astrocytoma. The postoperative EEG and fits improved and the interictal discharges disappeared. During a half year follow-up he remained seizure free.

The case is particularly interesting because the development of the focus contralateral to a temporal lobe tumour was detected first. However, the bitemporal involvement in the ictal pattern seemed to be evident and interictal epileptiform discharges appeared on the side of the tumour only later.

Which temporal lobe has the primary epileptic process can not be easily answered, and the term "mirror focus" seems meaningless, while bilateral contribution in the seizures is evident.

Based on these observations we assume that in many temporal lobe epilepsies seizures are determined by bilateral involvement of the temporo-limbic structures and the course of clinical events depends on the balance of the bilateral epileptic facilitatory or inhibitory influences. We therefore assume that in many temporal lobe epilepsies seizures are determined by bilateral involvement of the temporo-limbic structures and the course of clinical events depends on the balance of the bilateral epileptic facilitatory or inhibitory influences.

The interpretation of the beneficial influence of tem-

poral lobe resections as due to an excision of the epileptic focus can be questioned and—at least in certain cases—an other interpretation can be made namely, the disruption of the pathological bilateral limbic interplay.

From the practical point of view when intensive medical treatment is ineffective looking for an early surgical solution is highly reasonable.

The theoretical concept of secondary synchronization had been supported by early experimental data in animals (Ralston 1961) and by indirect evidences in epileptic patients (Tükel and Jasper 1952). The term "secondary synchronization" is principally used to characterize bilateral slow spike-wave synchronization assumed to project bilaterally by non-specific thalamo-cortical system triggered by chronic impulses of focal cortical origin, without obligatory clinical counterparts of the EEG phenomenon.

A large number of patients with intractable epilepsy show diverse seizures with partial and generalized features sharing the common characteristics of secondary

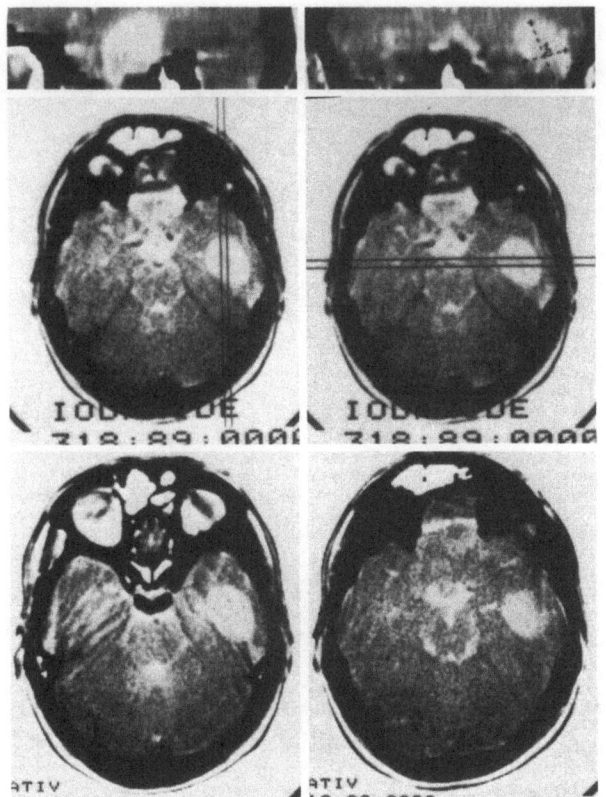

Fig. 4. Patient 3. CT scan delineating a right temporal cystic tumour

A 41-years-old male patient with late LGS showed great activation of repetitive fast discharges (rapid runs of spikes) during SWS. After anterior callosotomy the occurrence of the discharges where comparably the same as before surgery but the synchronization of the discharges were altered. The left and sometimes the right hemisphere showed a considerable latency in the involvement which parallelled a clinical improvement. The number and severity of harmful seizures have definitely decreased.

The same clinical improvement was detected in another young patient with LGS. The EEG showed unilateralization of the previously bihemispheral spike-wave paroxysms, a finding described in other patients (Spencer *et al.* 1987). However in another example contradictory results were demonstrated. In a 17-years-old girl with LGS the surgical intervention resulted in an extraordinary good clinical result. The seizures nearly ceased but the bihemispheral spike-wave synchronization remained absolutely the same as preoperatively.

Therefore we suggest that the existence or amount of bilateral EEG synchronization should not be regarded as a proper indicator of clinical failure or success in measuring or forecasting the results of callosotomy.

synchronized EEG with slow spike-waves. A search for a surgical solution in the form of callosotomy in these cases is nowadays widely accepted.

One of the main questions is whether the dissection or partial dissection of the corpus callosum can really stop that type of impulse traffic necessary for bihemispheral involvement in ictal epileptic excitation. The effect in our patients seems to verify this hypothesis.

References

1. Goldensohn ES (1984) The relevance of secondary epileptogenesis to the treatment of epilepsy: kindling and the mirror focus. Epilepsia 25 [Suppl] 2: S 156–S 168
2. Morrell F (1985) Secondary epileptogenesis in man. Arch Neurol 42: 318–335
3. Ralston BL (1961) Cingulate epilepsy and secondary bilateral synchrony. Electroenceph Clin Neurophysiol 13: 591–598
4. Spencer SS, Gates JR, Reeves AR, Spencer DD, Maxwell RE, Roberts D (1987) Corpus callosum section. In: Engel J (ed) Surgical treatment of the epilepsies. New York, pp 425–444
5. Tükel K, Jasper HH (1952) The EEG in parasagittal lesions. Electroenceph Clin Neurophysiol 4: 481

Correspondence : P. Halász, M.D., National Institute for Nervous and Mental Diseases, Budapest, Hungary.

Acta Neurochirurgica, Suppl. 46, 17–20 (1989)
© by Springer-Verlag 1989

Epilepsy Course in Cerebral Gangliogliomas: a Study of 16 Cases

M. Casazza[1], **G. Avanzini**[1], **G. Broggi**[2], **M. Fornari**[2], and **A. Franzini**[2]

Departments of [1] Neurophysiology and [2] Neurosurgery, Istituto Neurologico C. Besta, Milano, Italy

Summary

From November 1979 to March 1988, 16 patients with cerebral gangliogliomas were investigated and underwent surgery at the Institute Neurologico "C. Besta" of Milan. Their age varied from 11 to 48 years. 15 of these patients presented with a seizure as the first and often the only neurological symptom.

This report deals with the epileptologic and neuroradiologic features of these patients before and after surgical treatment.

Keywords: Epilepsy; ganglioglioma.

Introduction

Gangliogliomas are rather unfrequent tumours of the central nervous system consisting of mixed glial and nervous elements in different stages of differentiation with morphological neoplastic features[1-3].

After long discussions on their dysembriogenic or neoplastic nature, they are now recognized as slow growing tumours. They usually appear as a well circumscribed mass with cystic areas and frequently with a calcified component. The reported incidence varies from 0.6 to 0.9% of histologically diagnosed brain tumours, with higher incidence in infancy[4, 5].

Our series, collected during a 8.5-year period, confirm these data: 18 patients underwent surgery for ganglioglioma in our Institute, accounting for 0.5% of all the operated cerebral neoplasms and for 2.8% of paediatric age tumours.

Our purpose is to present the clinical and radiological features of these tumours and the outcome of the operated patients with particular interest in epilepsy, the most frequent presenting symptom of gangliogliomas.

Materials and Methods

Among 18 patients operated for gangliogliomas at the Istituto Neurologico C. Besta, Milan, from November 1979 to April 1988, we consider here 16 cases (9 males and 7 females) presenting with seizures as first symptom. The other two cases never had epileptic fits. We reviewed retrospectively the history of the epileptic patients, considered EEGraphic and neuroradiological data and neurologic examination before and at surgery. In 6 cases an Electrocorticographic (ECoG) intraoperative recording has been performed.

We successively followed the patients' outcome from clinical and neuroradiological point of view.

Results

Mean age of patients at operation was 23 ± 13 years (range 9 to 50); seizures set on in childhood (before 15 years) in 11 patients, in 3 of them they appeared even in the first year of life; the mean age of epilepsy onset was 13 ± 11.2 years (Table 1).

The mean interval between epilepsy onset and surgery was 10 ± 12 years, ranging from 1 to 40 years. This is due to very prolonged delays in performing neuroradiological exams in patients with few or no signs of suspect evolutive lesions.

Focal neurological signs were present at operation in 10 patients, they were mild and at least in 5 of them they appeared rather suddenly at a certain point of their clinical history, in probable coincidence with an abrupt increase of the cystic component of the tumour, that often led to more precise diagnostic investigations.

Increased intracranial pressure was observed in 2 patients, both with a very large cystic component of the tumour.

Psychic signs (mental retardation of different degree, behavioural and learning disturbances) were observed in the 4 patients with the earliest seizure onset (within the fourth year of life). In infancy 3 of these patients were diagnosed respectively as West syndrome, Lennox-Gastaut syndrome and post-vaccinic encephalitis with symptomatic partial epilepsy, according to clinical course and EEG data. Only later performed neuroradiological exams led to the correct diagnosis.

Table 1. *Gangliogliomas with Seizures as Presenting Symptom*

Case	Sex	Age at 1st seizure	Age at operation	Interval between onset and surgery	Focal neurological signs at operation	Increased intracranial pressure	Other signs
1. B. G.	f	18	22	4	—	—	—
2. B. G. L.	m	8	48	40	left hyperreflexia	—	—
3. B. I.	m	15	18	3	mild right hemiparesis	yes	—
4. C. C.	f	11	13	2	—	—	—
5. C. G.	m	13	17	4	right hyperreflexia + Babinski response	—	—
6. F. A.	m	1	22	21	—	—	moderate mental retardation
7. F. G.	f	4	9	5	—	yes	mild learning difficulties
8. G. A.	m	23	29	6	right hemianopia + hyperreflexia	—	—
9. G. D.	f	0.8	13	12	mild left hemiparesis	—	mild mental retardation
10. M. A.	f	19	20	1	mild right arm paresis	—	—
11. M. S.	m	16	17	1	—	—	—
12. R. E.	f	10	47	37	mild aphasia + right hemiparesis	—	—
13. S. A.	m	0.5	12	11.5	—	—	severe mental retardation + psychosis
14. S. D.	m	15	16	1	right hyperreflexia + Babinski response	—	—
15. U. M.	f	48	50 / 51 (reoperation)	2	mild right hemiparesis, hemianesthesia, hyperreflexia	—	—
16. Z. F.	m	7	15	8	mild right lower facial palsy + hyperreflexia	—	—

Table 2. *Postoperative Follow-up*

Case	Resection	Follow-up duration	Outcome	Recurrence
1.	total	3 years	seizure-free—no therapy	no
2.	total	1 year	moderately improved	no
3.	total	6.5 years	seizure-free—no therapy	no
4.	total	1 year	seizure free—with therapy	no
5.	total	3.5 years	seizure-free—no therapy	no
6.	total	1 year	seizure free—with therapy	no
7.	total	7.5 years	seizure-free—no therapy	yes
8.	total	2 years	improved	no
9.	total	2 years	improved	yes
10.	total	1.5 year	seizure-free with therapy	yes
11.	total	3 months	seizure-free with therapy	no
12.	total	5.5 years	improved	no
13.	subtotal	2 years	improved	yes
14.	total	1 year	improved	no
15.	1) total	1 year	improved	yes
	2) total	2 months	seizure-free	—
16.	subtotal	4.5 years	unchanged	yes

Surgery produced in these cases an improvement in cognitive functions and behaviour.

All our patients presented with simple partial or complex partial seizures, evolving in 4 cases to generalized tonico-clonic seizures. Seizure features appeared rather monomorphous in the clinical history of every patient. In most cases of very early onset further seizure types were present in childhood and later disappeared. In any case epilepsy presented with an unfavourable course in most patients, with progressive increase in seizures occurrence and severity and incomplete response to medical therapy.

Scalp EEG showed interictal focal slow and/or epileptiform activity in 15 patients, bitemporal nonspecific theta activity in 1 patient. Focal abnormalities were in accordance with seizure features.

The tumours were located in the temporal lobe (8 patients), parietal lobe (3 patients), fronto-parietal region (2 patients), frontal lobe, occipital lobe and parieto-temporal region (1 patient each). They often extended to neighbouring areas, so explaining the imperfect concordance between tumour location and EEG focality observed in some patients.

Diagnosis of possible evolutive lesion was made in all patients on the basis of CT scan and confirmed in 7 of them by MR.

Gangliogliomas appeared on CT scan either as low density enhancing lesions with only minimal mass effect (8 cases), or as cystic lesions with parenchymatous component and mass effect (7 cases), or as hypodense lesions with no enhancement (1 case). In 4 cases calcifications were observed; analogous images were present also on plain skull films.

In 2 patients a CT scan performed respectively 1 and 3 years before diagnosis was referred to as normal. In 5 patients a CT scan repeated at least 1 year after the first one showed signs of modest tumour growing.

Angiography was performed in 12 patients: in 7 it demonstrated only an avascular mass, while in 5 more cases an abnormal even if faint vascularization of the tumour was present.

All our patients underwent surgery (Table 2) with micro-operative technique: in 14 cases excision was apparently complete, in 2 cases it was subtotal. In 6 cases intraoperative ECoG was performed, but in only 3 cases the recording slightly modified the surgical strategy leading to an associated small cortical resection during tumour excision. Follow-up ranges from 3 months to 7.5 years.

Among the patients with total resection 4 are seizure-free without therapy, 4 are seizure-free with therapy (their follow-up is too short to withdraw antiepileptic drugs), 6 significantly improved and 5 of them present only with sporadic partial seizures or auras.

From the 2 patients who underwent subtotal excision 1 improved moderately and 1 is unchanged.

Recurrence was observed at the CT scan in 4 patients: only one of them has recently been reoperated, the others are periodically controlled since the CT finding has not yet clinical relevance.

Discussion and Conclusions

The high epileptogenicity of gangliogliomas has been repeatedly underlined by literature[4-9]. Our data referring to patients with very long-lasting histories of partial epilepsy stress the need of suspecting an evolutive, even if benign lesion, as ganglioglioma is, in every patient with partial epilepsy without known aetiology as shown by the cases previously diagnosed as idiopathic West and Lennox-Gastaut syndrome, even in spite of very early onset and of absence of focal neurological signs.

Surgery with complete removal of the lesion seems to be the most effective therapy in controlling epilepsy, considering the usual pharmacoresistance of these symptomatic epilepsies.

Careful preoperative neurophysiological studies either with scalp EEG or with stereo-EEG should probably be performed in order to define the epileptogenic area[10]. Nevertheless we think that ECoG can offer an additional tool in the surgical treatment of epileptogenic slow growing tumours as gangliogliomas, since the surgical flap planning has already been choosen on the basis of neuroradiological data.

In the last 6 operated patients we conducted intraoperative ECoG with the aim of determining the extension of the epileptogenic area and guiding the following excision. At present our data are still too scarce to evaluate the results of such technique in these cases.

Biopathological analysis of such tumours (*e.g.* cell kinetics) as well as detailed pre- and postoperative clinical and neurophysiological data may provide further information of practical interest for pathologists, neurosurgeons and mainly epileptologists.

References

1. Courville CB (1930) Ganglioglioma: Tumour of the central nervous system. Review of the literature and report of two cases. Arch Neurol Psychiatry 24: 439–491
2. Rubinstein LJ (1972) Tumours of the central nervous system. Armed Forces Institute of Pathology. Washington DC, pp 158–167
3. Zülch KJ (1965) Brain tumours: their biology and pathology, 2nd ed. Springer, Berlin Heidelberg New York, 184 pp
4. Demierre B, Sticnoth FA, Hori A, Spoerri O (1986) Intracerebral ganglioglioma. J Neurosurg 65: 177–182
5. Sutton LN, Packer RJ, Zimmermann RA, Bruce DA, Shut L (1987) Ganglioglioma in children. In: Homburger F (ed) Prog Exp Tumour Res 30. Karger, Basel, pp 239–246
6. Kalyan-Raman UP, Olivero WC (1987) Ganglioglioma: a correlative clinico-pathological and radiological study of ten surgically treated cases with follow-up. Neurosurgery 20-3: 428–433
7. Rossi E, Vaquero J, Martinez R, Garcia-Sola R, Bravo G (1984) Intracranial gangliogliomas. Acta Neurochir (Wien) 71: 255–261
8. Sutton LN, Packer RJ, Rorke LB, Bruce DA, Schut L (1983) Cerebral gangliogliomas during childhood. Neurosurgery 13: 124–128
9. Henry JM, Heffner RR, Earle KM (1978) Gangliogliomas of the CNS: a clinicopathological study of 50 cases (abstract). J Neuropathol Exp Neurol 37: 626
10. Munari C, Talairach J, Musolino A, Szikla G, Bancaud J, Chodkiewicz J (1983) Stereotactic methodology of "functional" neurosurgery in tumoral epileptic patients. Ital J Neurol Sci [Suppl] 2: 69–82

Correspondence: G. Broggi, M.D., Istituto Neurologico C. Besta, Via Celoria 11, I-20133 Milano, Italy.

Acta Neurochirurgica, Suppl. 46, 21–24 (1989)
© by Springer-Verlag 1989

Experience with Orbital Electrodes in the Patients Operated on for Epilepsy—Results of Temporofrontal Resections

I. Ribarić and **N. Sekulović**

Neurosurgical Clinic, University Clinical Centre, Belgrade, Yugoslavia

Summary

We introduced the orbital electrodes (Fb-frontobasal) in clinical practice in 1983. Since then we have operated on 112 patients for medically intractable "temporal" epileptic fits. In this series there were 45 patients (40%) with independent focal interictal EEG epileptic abnormalities over frontobasal cortex (with or without independent spiking over interomedial temporal region). In these patients we performed a small temporal lobectomy (with amygdaloectomy and resection of pes hippocampi) (TL) and resection of frontoorbital cortex (FOCR). The surgical results are as follows: There are 75 patients with follow-up one to 5 years. TL and FOCR were performed in 27 patients (36%). In this group seizure-free are 25 (93%) and 2 patients (7%) are much better. In the series of 48 patients examined by sphenoidal electrodes and operated for "temporal" fits (TL) by the same author (R. I.) (follow-up 5–13 years), seizure-free are 38 (79%), much better are 8 (17%) and unchanged are 2 (4%) patients. The difference in proportion of seizure-free patients in these two series is statistically significant ($t = 2.53$; $p < 0.05$).

Our results suggest that in some patients suffering with "temporal" seizures a well defined EEG picture suggests a new clinical entity the "temporo-frontal" epilepsy, the surgical treatment of which gives excellent results.

Keywords: Orbital electrodes, surgery for epilepsy; temporofrontal resection; temporo-frontal epilepsy.

Introduction

There is much evidence indicating anatomical connections of frontoorbital cortex with structures of limbic system[11,12,15] and temporal neocortex[23,25]. The term "orbito-insulo-temporopolar region", composed of the allocortical posterior orbital region and area temporopolaris (dysgranular cortex[3]), as well as the anterior insula, was introduced by Kaada in 1960[9] and its clinical implications were discussed by Gastaut and Lammers[7]. Some authors suggested that the frontobasal region might be involved in psychomotor epilepsy, as well as in psychotic schizophrenia-like states[1,2,22].

Long-term EEG monitoring with intracerebral depth electrodes gave us evidence of the involvement of the frontal lobe and particularly of the frontoorbital cortex in the epileptogenesis of complex partial seizures (CPS)[8,14,20,27].

Frontoorbital cortex is comparatively inaccessible to extracranial EEG recordings, although some information may be obtained with nasoethmoidal and supraorbital electrodes[13,19]. We introduced in clinical practice orbital (frontobasal-Fb) electrodes in 1983. The technique and the first clinical applications were reported in 1985[21].

Patients and Methods

In the last five years all candidates for surgery for medically intractable CPS in our Clinic were submitted to the following protocol for preoperative EEG evaluation: 1 to 3 standard 16-channel EEG recordings with international 10–20 position of scalp electrodes in awake patient (with 5 montages); 3 to 6 16-channel EEG recordings with sphenoidal (Sp) and orbital (frontobasal—Fb) electrodes in the awake patient and in pharmacologically induced sleep (Pentobarbital and Chlorpromazine), with 8 montages.

We identified a group of patients with unilateral epileptic interictal abnormalities being most prominent in the position of Fb electrode (phase reversal in Fb electrode). Very often a simultaneous spiking was recorded in Fb 1/2-T 3/4 (in Sp-Fb montages). In "sphenoidal montages" phase reversal were recorded in the Sp electrode. This type of EEG epileptic abnormalities we call "Fronto-Basal type" (FB type). In this group were not included the patients with equipotentiality (of interictal abnormalities) Sp = Fb or with maximum of the electrical field over the temporal base or convexity. Clinical epileptic manifestations in these patients were not specific.

In patients with unilateral FB-type of EEG abnormalities we performed a "small" temporal lobectomy (TL) with amygdaloectomy and resection of pes hippocampi (1 cm) and uncus, together with a resection of the posterior two thirds of frontoorbital cortex (FOCR). TL (in these patients) was restricted to the temporal pole ("temporal polectomy"—TP): posterior margin of TP starts at the upper margin of the temporal lobe, 2,5–3 cm behind the temporal pole. The resection plane slopes basoanteriorly to end at the inferior lobe margin 2–2,5 cm behind the temporal pole (Fig. 1). The posterior margin of FOCR starts 2–3 mm in front of the lateral end of the sphenoidal ridge and is directed vertically to the falx (we preserve the posteromesial "corner" of orbital cortex).

In this period of five years we have operated on 112 patients for medically intractable CPS. In this series there were 45 patients (40%) with FB-type of interictal epileptic abnormalities. They were submitted to TL (TP) and FOCR. The surgical results are as follows: There are 75 patients with follow-up 1 to 5 years operated on for intractable CPS. TL (TP) and FOCT were performed in 27 patients (36%). In this group (mean follow-up 2,5 years) seizure-free are 25 (93%) and 2 patients (7%) are much better (Table 1). In the series of 48 patients examined by sphenoidal electrodes and operated on for intractable CPS (TL) by the same author (I. R.) (follow-up 5 to 13 years) 38 (79%) are seizure-free, 8 (17%) much improved and 2 (4%) are unchanged (Table 2). The difference in proportion of seizure-free patients in these two series is statistically significant (t = 2,53; p < 0,05).

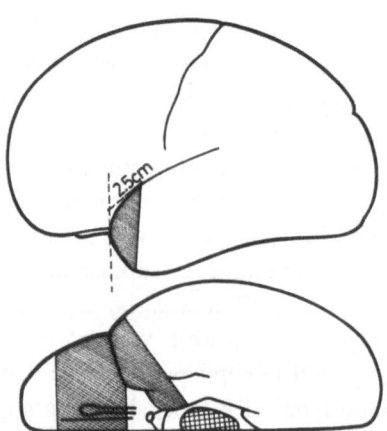

Fig. 1. The extent of resection in TP and FOCR (dashed area)

Table 1. *Surgical Results in "Temporo-Frontal EPI"* (operation: TL + FOCR)

Follow up (mean 2,5 years)	Seizure-free	Much better	Same	Total
1–2 years	9 (90%)	1 (10%)	0	10 (100%)
2–5 years	16 (94%)	1 (6%)	0	17 (100%)
Total	25 (93%)	2 (7%)	0	27 (100%)

Table 2. *Surgical Results in CPS Examined by Sp-Electrodes* (operation: TL)

Follow up	Seizure-free	Much better	Same	Total
5–13 years	38 (79%)	8 (17%)	2 (4%)	48 (100%)

Table 3. *Surgical Results in FOCR*

Seizure-free	Much better	Same	Seizure-free after reoperation (TP)	Total
2 (22%)	2 (22%)	1 (11%)	4 (45%)	9 (100%)
follow up 1,5 y			follow up 1–3,5 y (mean 1,8 y)	

In 9 patients with FB type of interictal epileptic abnormalities we performed only FOCR. The operative results are as follows (Table 3): Seizure-free are 2 patients (22%), much better are 2 (22%) and unchanged is 1 (11%) (follow-up 1,5 year). In 4 of these patients we performed TP one year after the first operation due to an unsatisfactory result of the first operation. These four patients are now seizure-free with follow-up of 1–3,5 years (mean 1,8 years). Before the first operation (FOCR) EEG evaluation in these patients showed FB type of interictal epileptic abnormalities and before the second operation (TP) EEG showed a maximum of the electrical field in the Sp electrode.

Electrocorticography was performed in 8 patients of the TL-FOCR series. In 6 cases some spiking was recorded over the cortex in the vicinity of the temporal pole, and independent, less frequent, spikes, sharp waves or slow sharp waves were recorded over the frontoorbital cortex. In 2 cases epileptic abnormalities were recorded only over the temporal pole cortex.

In all cases of TL-FOCR and FOCR series signs of neuronal degenerations and gliosis were present in the frontoorbital cortex. There were also changes compatible with "mesial temporal sclerosis"[6]. In five of these cases there were small benign gliomas in lateral or mesial temporopolar regions.

Discussion and Conclusions

Our results of TL (TP) and FOCR, although with comparatively short follow-up, appears to be more favourable than those reported in the literature[5, 24]. It should also be noted that the temporal resections in this series were comparatively small. Our experience with TL shows that the seizures are more successfully abolished or reduced by larger TL as has also been emphasized by others[4, 5, 18, 26].

There are two possible explanations for the good results in this series: 1. FB type of interictal epileptic abnormalities seems to be a useful base for selecting the patients for either TP or FOCR; 2. The simultaneous TP and FOCR procedures in the patients selected on the basis of FB type of interictal epileptic abnormalities appear to be particularly suitable.

Experience with supraorbital and ethmoid electrodes showed that in some cases it was possible to record discharges from the temporopolar region[10]. The same could be expected to occur with orbital electrodes. The difference between EEG findings before the first and the second operations in four reoperated patients suggest that FB type of interictal epileptic abnormalities recorded before the first operation did not originate only from temporopolar region. Our electrocorticographic findings shows that in some of these cases selected on the basis of FB type of EEG abnormalities there are independent epileptic discharges in frontoorbital and temporopolar cortex. In view of the temporo-frontoorbital connections one could consider these two cortical regions to be one "epileptogenic area". It seems logical that in these cases (with temporoorbital spiking) the best chances for elimination of intractable epileptic seizures is obtained when the "epileptogenic area" (TP and FOCR) is completely removed.

On the basis of the reported experience, it seems to us justifiable to regard patients with FB type of interictal epileptic abnormalities as a group of "temporofrontal epilepsy", which may benefit from a standardized surgical procedure. It can be concluded: 1. The examination of the patients is comparatively simple, inexpansive harmless and the results are reliable; chronic intracranial recording is not necessary; 2. The resection does not comprise the most sensitive structures, the posterior hippocampus and temporal convexity, which are functionally important for memory and language[16]; 3. The surgical technique is standardized and comparatively simple. Electrocorticography and functional mapping are not necessary; 4. The outcome of surgical treatment is rewarding.

References

1. Ajuriaguerra JD, Tissot R (1977) Rhinencéphale: Neurotransmitters et psychoses. Georg SA, Genève, Masson, Paris
2. Akert K (1980) Anatomische und physiologische Grundlagen zum Problem der psychomotorischen Epilepsien und des Status psychomotoricus. In: Karbowski (ed) Status psychomotoricus and seine Differentialdiagnose. Hans Huber, Bern, pp 9–38
3. Bailey P, Bonin GV (1951) The isocortex of man. University Illinois Press, Urbana
4. Crandall PH, Walter RD, Rand RW (1963) Clinical application of studies on stereotactically implanted electrodes in temporal lobe epilepsy. J Neurosurg 20: 827–840
5. Falconer MA (1976) Anterior temporal lobectomy for epilepsy. In: Smith CR (ed) Operative surgery, vol 14. Butterworths, London, pp 142–149
6. Falconer MA, Serafetinides EA, Corsallis JAN (1964) Etiology and pathogenesis of temporal lobe epilepsy. Arch Neurol 10: 233–248
7. Gastaut H, Lammers HJ (1961) Anatomie du rhinencéphale. In: Alajouanine T (ed) Les grande activités durhinencéphale. Masson, Paris, pp 1–166
8. Geir S, Bancaud J, Talairach J, Bonis A, Szikla G, Enjelvin M (1977) The seizures of frontal lobe epilepsy. Neurology 27: 951–958
9. Kaada BR (1960) Cingulate, posterior orbital, anterior insular and temporal polar cortex. In: Field J et al (eds) Handbook of physiology, sect 1, vol II: Neurophysiology. Williams and Wilkins, Baltimore, pp 1345–1372
10. Lesser PR, Lüders H, Morris HH, Dinner SD, Wylie E (1987) Commentary: extracranial EEG evaluation. In: Engel J (ed) Surgical treatment of the epilepsies. Raven Press, New York, pp 173–179
11. Lindvall O, Björklund A (1974) The organization of ascending catecholamine neuron systems in the rat brain. Acta Physiol Scand 412: 1–48
12. Livingston KE, Excobar A (1971) Anatomical bias of the limbic system concept. A proposed reorientation. Arch Neurol (Chicago) 24: 17–21
13. Morris HH, Lueders H (1985) Electrodes. In: Gorman J, Ives JR, Gloor P (eds) Long-term monitoring in epilepsy. Electroencephalogr Clin Neurophysiol [Suppl 37]. Elsevier, Amsterdam
14. Munari C, Stoffels C, Bossi L, Bonis A, Talairach J, Bancaud J (1981) Automatic activities during frontal and temporal lobe seizures: are they the same? In: Dam M, Gram L, Penry KJ (eds) Advances in epileptology: XIIth Epilepsy International Symposium. Raven Press, New York, pp 287–291
15. Nauta NJH (1979) Expanding borders of the limbic system concept. In: Rasmussen T, Morino Jr R (eds) Functional neurosurgery. Raven Press, New York, pp 7–23
16. Ojeman AG, Engel Jr J (1987) Acute and chronic intracranial recording and stimulation. In: Engel Jr J (ed) Surgical treatment of the epilepsies. Raven Press, New York, pp 263–288

17. Olivier A (1987) Commentary: cortical resection. In: Engel Jr J (ed) Surgical treatment of the epilepsies. Raven Press, New York, pp 405–416

18. Penfield W, Baldwin M (1952) Temporal lobe seizures and the technique of subtotal temporal lobectomy. Ann Surg 136: 625–634

19. Quesney LF (1987) Extracranial EEG evaluation. In: Engel Jr J (ed) Surgical treatment of epilepsies. Raven Press, New York, pp 129–166

20. Quesney LF, Krieger C, Leitner C, Gloor P, Olivier A (1984) Frontal lobe epilepsy: Clinical and electrographic presentation. In: Porter RJ, Mattson RH, Ward AA, Dam M (eds) Advances in epileptology: XVth Epilepsy International Symposium. Raven Press, New York, pp 503–508

21. Ribarić I, Sekulović N, Stefanović B (1985) Frontobasal transorbital electrodes—new technique and clinical experience with 30 operated on and 40 non-operated epileptic patients. 8th International Congress of Neurological Surgery, Toronto, Abstracts, p 33

22. Stevens JR (1973) Psychomotor epilepsy and schizophrenia: a common anatomy? In: Brazier MAB (ed) Epilepsy, its phenomena in man. Academic Press, New York, pp 191–214

23. Truex CR, Carpenter BM (1973) Human neuroanatomy, 6th ed. William and Wilkins, Baltimore

24. Van Buren MJ, Ajmone-Marsan L, Mutsuga N, Sadowsky D (1975) Surgery of temporal lobe epilepsy. In: Purpura DP, Penry KJ, Walter DR (eds) Advances in neurology, vol 8: Neurosurgical management of the epilepsies. Raven Press, New York, pp 155–196

25. Van Hoesen GW, Mesulam MH, Haaxma R (1976) Temporal cortical projections to the olfactory tubercle in the rhesus monkey. Brain Res 109: 375–381

26. Walker AE (1967) Temporal lobectomy. J Neurosurg 26: 642–649

27. Wieser HG (1983) Electroclinical features of the psychomotor seizure. Gustav Fischer, Stuttgart London New York

Correspondence: I. Ribarić, M.D., Neurosurgical Clinic, University Clinical Centre, Belgrade, Yugoslavia.

Acta Neurochirurgica, Suppl. 46, 25–27 (1989)
© by Springer-Verlag 1989

Design of an Intracranial Electrode for Monitoring

V. M. Garcia Marin, M. J. Rodriguez Palmero, L. Gonzalez Feria, M. Ginoves Sierra, D. Martel Barth-Hansen, and **J. Ravina Cabrera**

Department of Neurosurgery, University Hospital of Canary Islands, Medicine Faculty, La Laguna, Tenerife

Summary

Direct recording of the electric brain activity gives more information than conventional electroencephalogram. Several authors have designed a variety of electrodes in order to solve the different problems of electrocorticography but in our opinion none of these fulfil the following features: easy implantation and extraction with minimal trauma; flexibility to allow placement over regions of the brain cortex that are difficult to access (interhemispheric fissure, medial aspect of the temporal lobe, frontobasal region, etc.), good quality recording.

A multiple contact electrode which we think matches these features has been designed. Initially this electrode was tested in the postoperative monitoring of ten patients with supratentorial malignant tumours and in one case of intractable epilepsy.

In two patients complications of treatment were detected, one had an epileptic seizure and the other had bleeding in the tumoural bed. In the first case a right temporal focus was delineated and posteriorly excised.

Implantation of the electrode was always very simple, either from the craniotomy or from a burr hole, and its flexibility allowed us to place it over the regions above mentioned. Also the extraction was easy with a simple traction and without the need for a second intervention. In all cases the recording quality was excellent.

Keywords: Electrocorticography; epilepsy; intracranial electrodes; monitorization.

Introduction

Direct recording of electric brain activity gives more information than scalp recording. Besides the elimination of noise of the conventional electroencephalogram, it permits the study of medial and basal cortex of the hemispheres. Moreover, the possibility of obtaining records with very small interpolar distance enables more accurate localization[1, 2, 7].

As far as the position of the electrodes is concerned, there are two main groups of these: cortical electrodes and intracerebral ones[3–6, 8, 9]. Stereotactically implanted deep electrodes have been widely used for recording activity both from the cortex and from deep gray matter[7]. Goldring et al.[3, 4] have used epidural electrodes. In 1986, Rosembaum et al.[6] reported the use of a cylindrical electrode with 5 contacts, made with a silicone tube and 5 stainless steel wires rolled around it. The one we present here is very similar. It is placed through a burr-hole and can be removed by simple traction. Nevertheless, Rosembaum found some difficulties to place it over certain regions of the brain, for example, in the frontal lobes.

In our opinion, none of these electrodes fulfill the following features: easy implantation and extraction, minimal trauma and flexibility to placement over areas of cortex difficult of access, and good recording qualities. We present a multiple contact electrode for cortical recording from the subdural space which fulfill the above features.

Methods and Material

The electrode is made on a silicone tube, 20 cm long, 2 mm external diameter and a 1 mm wide lumen. The electric contacts are made from stainless steel rings, 2 mm long, placed 10 mm apart. The number of contacts varies from 3 to 10, according to the need. The contact surface of each cylinder is 12.5 mm^2 and each is connected to teflon isolated stainless steel wire, 0.03 mm in diameter. The wires run within the silicone tube and end in a connector adapted to the EEG pre-amplifier (Fig. 1). Sterilization of the electrode is carried out in ethylene oxide.

Clinical testing was done by monitoring the postoperative development in 10 patients operated upon for supratentorial malignant brain tumours and for localization in an epileptic focus. The electrode was inserted through a burr-hole of the craniotomy in the cases of brain tumours. In the case of intractable epilepsy a single burr-hole was made for the implantation. In all cases, the electrode emerged 5 cm away from the main incision.

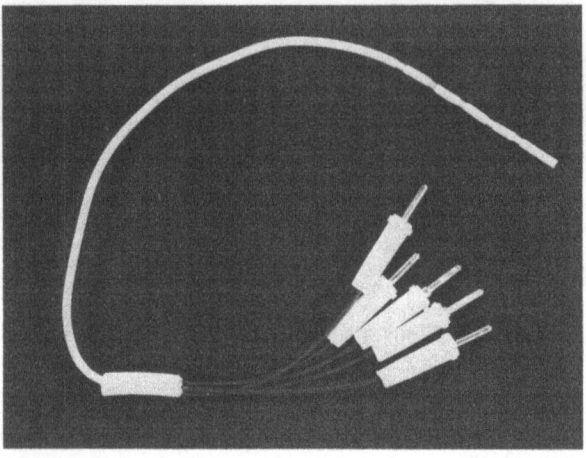

Fig. 1. Photograph of the electrode. In this case it has five contacts seen on the top-right side of the image

Results

The quality of the electrocorticographic recordings was excellent, the impedances being less than 7 kohm. The electrodes were easily placed over comparatively inaccessible regions of cortex, like medial and basal aspects of the lobes (Fig. 2). The electrodes were kept in place from 2 to 7 days, with an average of 4.1 days. The extraction, by means of a simple traction, was always easy. No re-operation was needed. There were no complications. In one case, considerable spike activity was detected over the operated region 24 hours before the appearance of epileptic fits. In the patient with intractable epilepsy, a right temporal epileptic focus was found (Fig. 3).

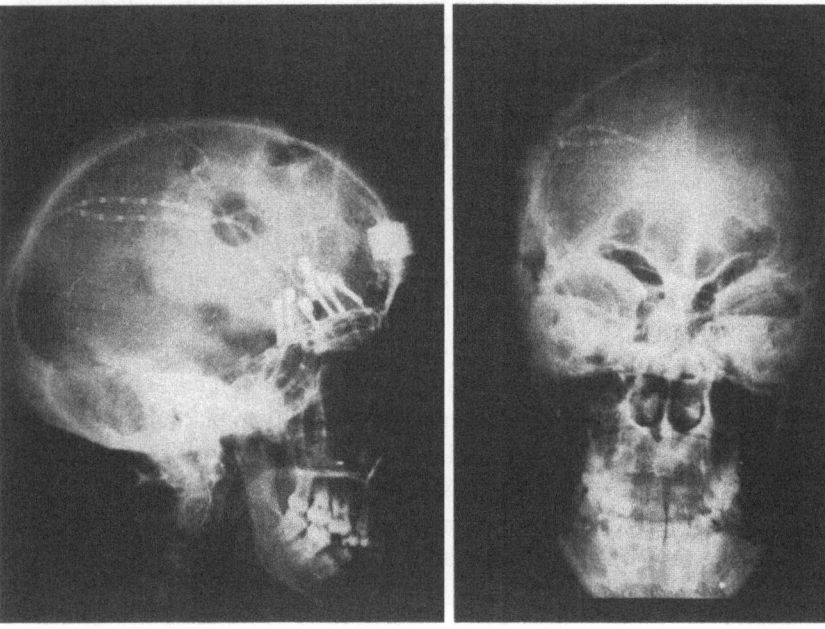

Fig. 2. Calvarium radiograph showing three different electrodes placed in a patient: two over right parieto-occipital region and the third around the right temporal lobe

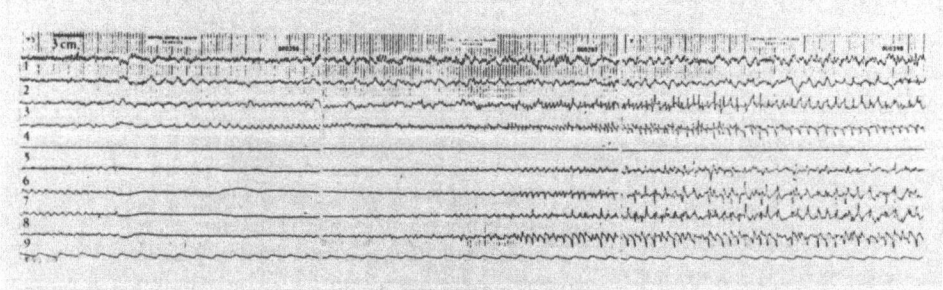

Fig. 3. Electrocorticogram of the patient with intractable epilepsy, showing a fit beginning in the right temporal lobe (Channels 5 to 8, both included)

Discussion

From our experience we can say that our electrode has some advantages compared to other subdural electrocorticographic electrodes. It is easier to place over distant zones than Wyler and co-workers' electrode[9], and re-operation is unnecessary.

It is very similar to the electrode desgined by Rosembaum *et al.*[6], although our larger number of contacts allows no cover up to 10 cm of brain cortex.

We are trying to reduce the external diameter of the electrode to further diminish the trauma of implantation and extraction. Indications for electrocorticography could be applied not only to intractable epilepsy but to monitoring the postoperative course of other neurosurgical conditions.

References

1. Crandall PH (1973) Developments in direct recordings from epileptogenic regions in the surgical treatment of partial epilepsies. In: Brazier MAB (ed) Epilepsy: its phenomena in man. Academic Press, New York, pp 288–310
2. Crandall PH, Walter RD, Rand RW (1963) Clinical applications of studies on stereotactically implanted electrodes in temporal lobe epilepsy. J Neurosurg 21: 827–840
3. Goldring S (1978) A method for surgical management of focal epilepsy, specially as it relates to children. J Neurosurg 49: 344–356
4. Goldring S, Gregorie EM (1984) Surgical management of epilepsy using epidural recordings to localize the seizure focus. Review of 100 cases. J Neurosurg 60: 457–466
5. Maxwell RE, Gates JR, Fiol ME, Johnson MJ, Yap JC, Leppik IE, Gumnit RJ (1983) Clinical evaluation of a depth electroencephalography electrode. Neurosurgery 12 (5): 561–564
6. Rosembaum TJ, Laxer KD, Vessely M, Smith WB (1986) Subdural electrodes for seizure focus localization. Neurosurgery 19 (1): 73–81
7. Talairach J, Bancaud J (1974) Stereotaxic exploration and therapy in epilepsy. In: Vinken PJ, Bruyn GW (eds) Handbook of clinical neurology, vol 15. North Holland, Amsterdam, pp 758–782
8. Wieser HG, Elger CE, Stodieck SRG (1985) The foramen ovale electrode: A new recording method for the preoperative evaluation of patients suffering from mesio-basal temporal lobe epilepsy. Electroencephalogr Clin Neurophysiol 61: 314–322
9. Wyler AR, Ojeman GA, Lettich E, Ward Jr AA (1984) Subdural strip electrodes for localizing epileptogenic foci. J Neurosurg 60: 1195–1200

Correspondence: V. M. Garcia Marin, M.D., Department of Neurosurgery, University Hospital of Canary Islands, La Laguna 38230, Tenerife, Spain.

Spasticity and Movement Disorders

We extensively studied the relationship between structures and muscles, dentate nucleus and muscles ipsilaterally and P, VL, Vim, CM, motor cortex and dentate nucleus contralaterally.

Materials and Methods

Our investigations were carried out on 78 patients; 57 had motor disturbances (parkinsonism, choreo-athetosis, atactic tremor, ballismus), 6 patients had intractable pain (phantom pain, thalamic pain, anaesthesia dolorosa) and 15 patients suffered from epilepsy.

Our chronic multielectrode method was published in detail elsewhere[17, 23, 25].

Results

The interpretation of the different parts of the elicited event from their projections, evoked potentials and motor modulation is summarized on Fig. 1. There are at least three distinguishable parts of the evoked potentials and motor modulation reflecting functionally different periods of the elicited events. The different periods of the elicited events within the motor system

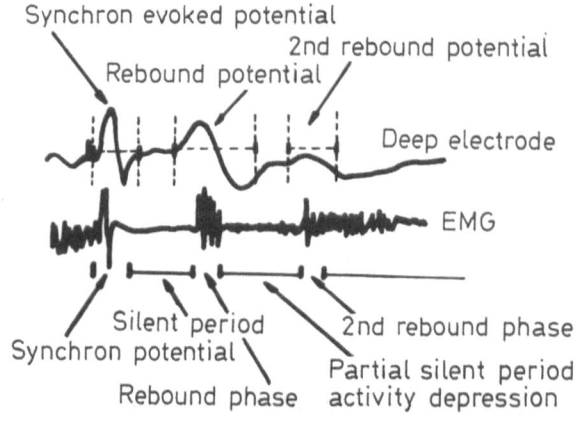

		Evoked potential	Motor modulation
First part	Spreading of irritation within the structures	The first spike (steep)	Synchron or asynchron potentials
Second part	Depression of the disturbed activity	Transitional activity	Silent period
	Rebound of activity	Second spike (large, broad)	Asynchron group of the enhanced activity
Third part	Damped oscillation in the control circuits, the original activity restored	Incomplete second and third parts if any	

Fig. 1.

parallel the different parts of evoked potentials and motor modulation and relate to normal function. This must be true for the pathological motor function as well.

The first part of the evoked potential and motor modulation represents the extension of the excitation on the functioning pathways, and thus somewhat represents the functional anatomy. The second part represents the supression of the disturbed function and the third part represents the elimination of the transient effect, reorganization and a rebound according to the originally intended function. Therefore, the third part essentially represents the functional characteristics of a given system. Naturally, we collected most of the pathological signs in this third part which was also most interesting from the point of view of the normal motor system.

The expression of the different periods of the elicited event and its sensitivity to the ongoing motor function depends on the site of the stimulation. To study this sensitivity we investigated evoked potentials and motor modulation during a simple motor event, which means, during preactivity rest, increasing contraction of a given muscle to a certain level, sustained contraction and during contraction release. There are two main types of sensitivity during a simple motor event (Fig. 2). To illustrate the positive sensitivity to stimulation of the pallidum or thalamus (VL, Vim, CM) recording of the evoked potentials in the motor cortex is appropriate, with stimulus parameters 0,005–0,5 ms square waves and amplitudes insufficient to evoke motor responses at rest on the appropriate side (Fig. 3 A, B). During preactivity rest the evoked potentials in the motor cortex are, usually, first spike, transitory wave, second spike, sometimes a few oscillations without motor reaction in the contralateral muscles.

During increasing motor activity in a choosen muscle, the evoked potentials become shorter and the sec-

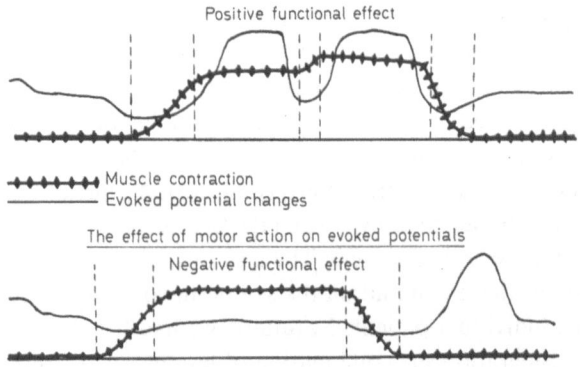

Fig. 2.

Acta Neurochirurgica, Suppl. 46, 31–36 (1989)
© by Springer-Verlag 1989

Some Basic Features of the Pathological and Normal Motor System Studied by Chronic Deep Electrodes

Sz. Tóth[1], Z. Tóth[2], J. Vajda[3], and A. Sólyom[3]

[1] University of Debrecen, Medical School Neurosurgical Clinic, Debrecen, Hungary, [2] MAV Central Hospital, Neurological and Neurosurgical Department, Budapest, Hungary, [3] National Institute of Neurosurgery, Budapest, Hungary

Summary

Our stereotactic experiences in agreement with the literature showed, that different target points could influence the same motor disturbance. To choose the best target point or target point combinations we generally implant the electrodes into VL, Vim, CM, P, dentate nucleus and motor cortex. To ensure the correct sequence of therapeutic lesions we developed an investigation system, taking into consideration the resting and the working state of the motor system.

We elicited events centrally (stimulation of the different target points) and peripherally (reflexes) and recorded the evoked potentials at the non-stimulated sites along with the motor and motor modulation effects in the appropriate muscles.

The elicited events depend on the site of stimulation and registration and on the state of muscle activity. The centrally and peripherally elicited events influence each other. With our technique the elicited events and their functional dependency is most explicit within the motor system. The results help to explain some basic motor functions and help to answer some of our therapeutic questions.

Keywords: Chronic electrode; deep structures (basal ganglia, cerebellum); motor diseases; motor function.

Introduction

Our stereotactic experiences in agreement with the literature showed that different target points in the motor system could influence the same motor disturbance[1, 3, 5, 14–16, 23–25]. During the history of surgery for movement disorders lesions in the pallidum, ventrolateral, ventral intermedial, dentate nuclei and centrum medianum were all effective as well as lesions in the pyramidal system. The result will depend on the site of the lesion and on the symptom combination of the individual patient. Naturally, we had to take into consideration the combination of lesions for the most effective therapy. Therefore we implanted chronic deep electrodes into different target points (P, VL, Vim, CM, dent. nucl., motor cortex) and created a system of investigations for choosing the best target point or combination of target points for the symptom complex of a given patient[25].

The investigation consisted of the study of elicited events centrally, by means of electrical stimulation of the different target points and peripherally, by means of myotatic reflexes elicited by electrical or mechanical stimulation. The registration was carried out according to the elicited events: evoked potentials, at the non-stimulated target points, during motor modulation, in the contracting muscle. We studied these events at rest, without muscle activity and also during voluntary contraction, with chronically implanted deep electrodes and concentric needle electrodes (EMG[19–24, 26]). Within the motor system, stimuli either central or peripheral, elicit a series of events, which can be followed in their projections either as evoked potentials at the non-stimulated points or as motor modulation in the appropriate contracted muscle. The elicited series of events depends on the site of stimulation and on the normal or pathological state of the stimulated system. The projections of the elicited series of events, namely the evoked potentials and motor modulation will depend on the site of stimulation and on the site of registration along with their controlling circuits which are working according to the state or motor activity (rest or contraction).

According to our system of investigation a structure can be defined as motor either when one can elicit a motor or motor modulation effect from it or when the registrated evoked potentials from it are under the influence of motor activity.

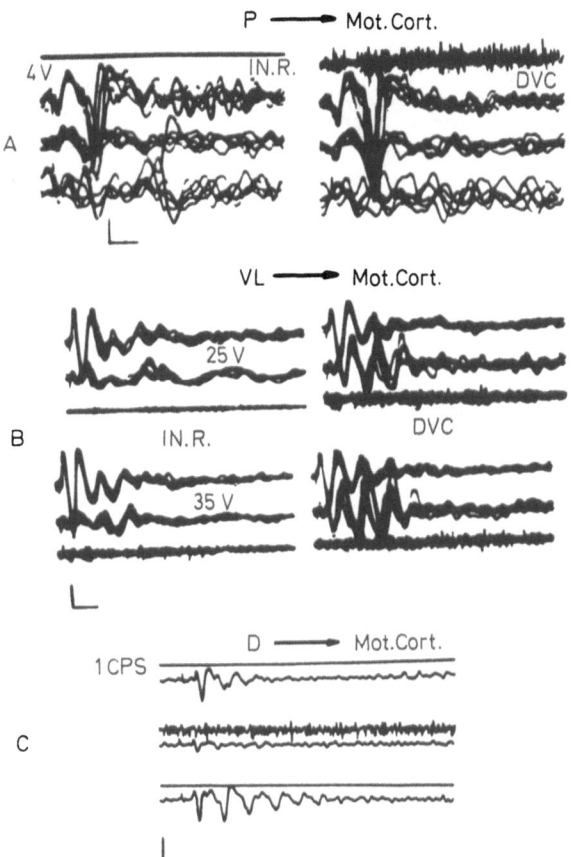

Fig. 3. A) The evoked potential changes of the right motor cortex during stimulation of the right pallidum, at rest and during voluntary contraction of the left biceps muscle. Calibration: 50 ms, for muscle 1,000 microV, for the 2nd and 4th channel 50 microV, for the 3rd channel 200 microV. Photosuperposition technique. Parkinsonism. The effect is definitely positive (enhancing). B) Evoked responses in the left motor cortex during stimulation of the left VL with 0,05 ms, 1 cps square stimuli of increasing voltage, during rest (IN.R) and during voluntary contraction (DVC), of the right biceps muscle. Calibration: 1st channel 0,1 mV/cm, second channel 0,05 mV/cm and 3rd channel 1 mV/cm, time calibration 50 ms. Parkinsonism. Definitely positive functional effect. C) Effect of stimulation of the dentate nucleus on the motor cortex and on the activity of the biceps muscle at rest and during voluntary contraction. No motor modulation effect. A definitely negative functional effect. In the postactivity period there is a rebound like increase of the evoked potential in the motor cortex. Calibration: 30 microV, 50 ms, for muscle 2,1 mV/cm, stimulus parameters 0,07 ms, 30 V, 1 Hz Parkinsonism

ond spike smaller, and immediate synchronous potential appears in the muscle, according to the first spike of the evoked potential without a silent period. During sustained contraction the second spike in the elongated evoked potential becomes higher as in rest (rebound). In the muscle activity there is an explicit silent period after the synchronous potential and according to the increased second spike in the evoked potential there is an expressed grouped rebound activity in the muscle.

Contraction release is similar during increasing activity and postactivity relaxation as in preactivity rest. By stimulating P, CM, and motor cortex the first part of the evoked potential and motor modulation is more obvious, while stimulating VL and Vim the second (rebound) part is more obvious.

To illustrate the negative sensitivity stimulation of the dentate nucleus recording in the opposite motor cortex and a choosen muscle on the same side as the dentate nucleus is appropriate (Fig. 3 C). During voluntary contraction on the side of the dentate nucleus the evoked potentials decrease (the first spike as well as the rebound part) and they remain low during the whole activity period. In postactivity rest, for a few seconds the evoked potentials become greater and longer as in preactivity rest. With stimulation of the dentate nucleus (cerebellum) by single stimuli there is no motor or motor modulation response, but by double stimuli (2–4 ms interval) motor modulation begins with a silent period.

What do these results suggest for special functions of the different structures? Motor cortex, pallidum, CM are working with a fairly direct connection to the contralateral muscles. They are involved in quick movements, mobilizing control circuits but mostly for stabilizing predetermined short movements. VL and Vim are work with strong feed back control; they have a tendency for oscillation (normally damped) and they take part in and control long-lasting, sustained movements. Dentate nucleus (cerebellum) has a characteristic of suppressing the strong feedback control which smoothes the movements. We do not understand the real mechanism but the cerebellum must have an immediate important postactivity function.

The central and peripherally elicited events acting through the same system can interfere, depending on the interval of the two stimulation (Fig. 5).

In parkinsonism, the third part of the evoked potentials is more pronounced as in other disorders. In the motor modulation the rebound activity following the silent period is strong. The reactive rebound phase elicited oscillation is not well damped as is well illustrated in the evoked potential and motor modulation (Figs. 3 B and 4 A). Whilst sustained tremor-like oscillation with appropriate stimulus parameters are easy to produce[26].

In choreo-athetosis the first part of the evoked potential is well pronounced but the reactive third part is not well developed. The same is true for motor modulation. This corresponds to the pathological movement as the innervation is going from one muscle to

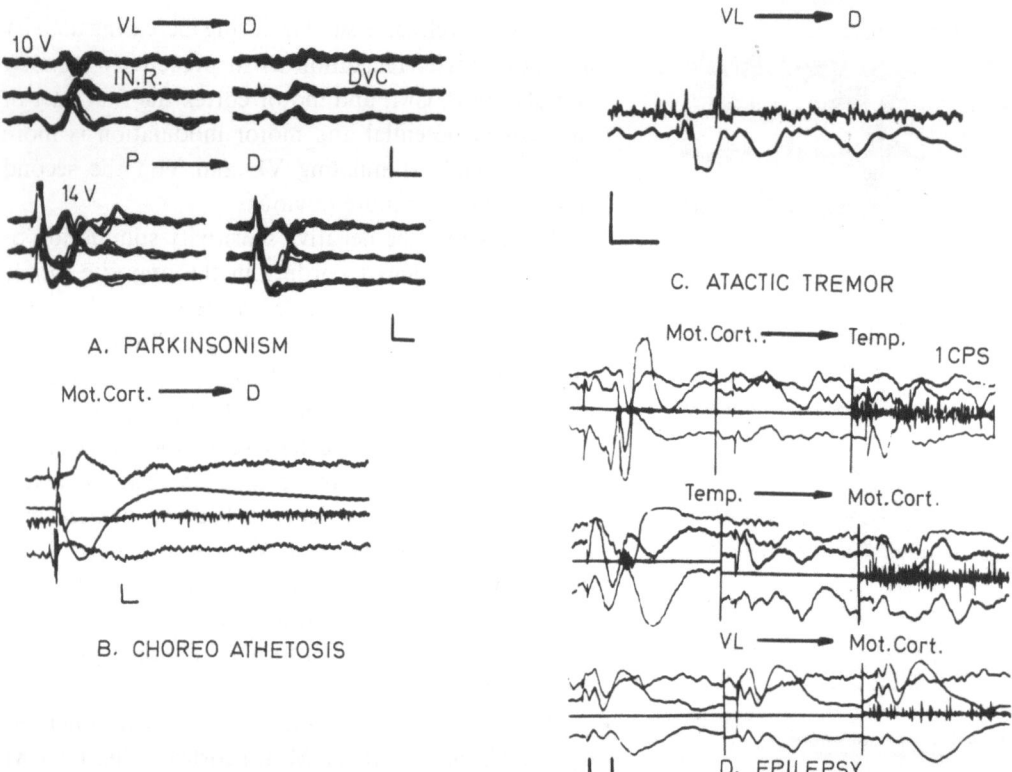

Fig. 4. A) Parkinsonism. Change of evoked responses in the left dentate nucleus during stimulation of the right VL and pallidum respectively, with 0,05 ms, 1 cps square stimuli at rest and during voluntary contraction of the left biceps muscle. Calibration: 50 ms, for all channels 100 microV. B) Choreo athetosis. Evoked potential from the left dentate nucleus, while stimulating the right motor cortex, during voluntary innervation of the left biceps muscle. During contraction the synchronous action potential definitely increases. After silent period there is no motor rebound, there is a tonic increase, as we can see on the mechanogram. Calibration: 1–4 channels 0,05 mV/cm, channel 2 0,5 mV/cm, mechanogram 2 kg/cm. Stimulus parameters: 0,05 ms, 50 V. Time 100 ms. C) Atactic tremor. Evoked potential in the left dentate nucleus while stimulating the right VL, during voluntary innervation of the left biceps muscle. Calibration: for muscle 2,1 mV/cm, for D 30 microV/cm. Stimulus parameters: 0,07 ms, 40 V. Time 63 ms/cm. D) Epilepsy. Epileptic spikes can be registered from all of the registration points; it can be evoked from all of the stimulated points by the first stimulus and it is accompanied by an asynchronous potential group as motor answer in the registrated muscle. The second column shows that during rest the stimulation does not evoke epileptic activity. The third column shows that motor modulation effect in a contracting muscle can be evoked but only from stimulating points inside the motor system. Calibration: channel 1, 2, 4 10 microV/cm, channel 3 500 microV/cm. Stimulus parameters: 0,05–0,2 ms. 25–50 V. Time 100 ms

the other continuously with very little feed back control (Fig. 4 B).

The rebound activity in the motor modulation after the silent period is usually an asynchronous potential group but in atactic tremor the rebound activity is a large synchronous potential often larger then the first presilent synchronous potential, due to the lack of the cerebellar inhibition or damping effect (Fig. 4 C).

In epilepsy when an epileptic spike appears after stimulation this appears in the third phase of the evoked potential parallel with grouped potentials in the contralateral muscle. One can elicit motor reactions with epileptic spikes from outside of the motor system. In the elicited series of events one can detect the rhythmic properties of the particular epilepsy (Fig. 4 D).

Discussion

The significance of stereotactic surgical procedures in clarifying the functions of the motor system can not be over emphasized. It is justifiable to claim that the results of surgical operations and the intraoperative examinations, particularly with essential chronic deep electrodes, have permitted a new interpretation of motor function. Through this, it is possible to study the laws governing the motor system[9, 14, 16–21, 23, 24, 26]. In complex investigations, certain parts of the evoked potential might suggest where the changes in the spinal motor mechanism can be found. This enables a detailed study of how the central mechanism alters the relatively simple spinal reflex state, how the loss of certain motor

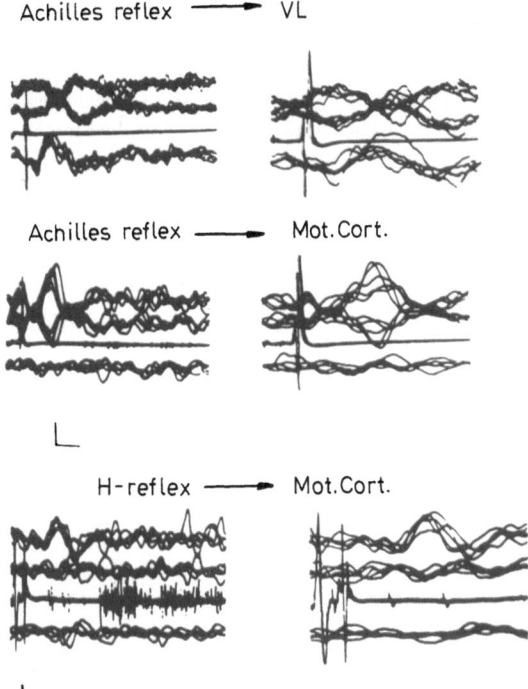

Fig. 5. Parkinsonism. Evoked potentials from the left VL and motor cortex, elicited by an Achilles reflex, and from the left motor cortex elicited by a H-reflex. Photosuperposition technique. Calibration for the Achilles reflex: 1, 2, 4 channel 0,1 mV/cm channel 3 1 mV/cm. Stimulus parameters; 0,05 ms, 65 V, time 50 ms. Calibration for the H-reflex: Channel 1, 2, 4 0,1 mV/cm, channel 3 0,5 mV/cm. Time 50 ms/cm. Stimulus parameters: 0,05 ms. 80 V

structures influence the evoked reflex mechanism and how it changes the evoked reflex mechanism if it is elicited from different points of the motor system. There is the possibility of revealing the type of changes that can be produced by motor action in the reflex-like responses of the motor system, or its reverse action. These correlation help us to clarify the connections already presumed by several authors in the form of long loop reflex paths between the spinal reflex mechanism and the more central motor system and makes a bridge between central motor system investigations and segmental spinal motor investigations[2–13, 16, 18–22, 24].

The response to the stimulation of the motor system is never a single event but a series of events reflecting the control process irrespective of where the stimulus affects the motor system and whether it is independent of any direct anatomical connection between the sites of stimulation and the recording sites.

The information obtained by complex examinations of the motor system greatly promotes the evaluation of the results of clinically available methods and last but not least helps to achieve better surgical results.

References

1. Bertrand C, Martinez N, Hardy J (1966) Localization of lesions in parkinsonism. J Neurosurg 24: 446–448
2. Burg DA, Struppler A, Szumsky AJ, Velhov F (1973) Generalized activation of the gamma motor system during restricted voluntary muscle activity. Electroenceph Clin Neurophysiol 34: 826
3. Divitiis E de, Giaquinto S, Signorelli CD (1971) Peripheral influences on VPM-VPL thalamic nuclei in the human. Confin Neurol 33: 174–185
4. Gassel MM (1970) A critical review of evidence concerning long-loop reflexes excited by muscle afferents in man. J Neurol Neurosurg Psychiatry 33: 358–362
5. Gillingham FJ (1960) Surgical management of the dyskinesias. J Neurol Neurosurg Psychiatry 23: 347–348
6. Hufschmidt HJ (1954) Die rasche Willkürcontraction. Z Biol 107: 1-24
7. Lee R, Withe G (1973) Modification of somatosensory evoked responses by voluntary movement. Electroenceph Clin Neurophysiol 34: 706–707
8. Magladery JW, McDougall DS (1950) Electrophysiological studies of nerve and reflex activity of normal man. Bull Johns Hopkins Hosp 86: 265–290
9. Marossero F, Cabrini GP, Ettore G, Infuso L (1972) Electromyographic study of motor responses following electrical stimulation of the corticospinal tract in man during stereotaxy. Confin Neurol 43: 230–236
10. Monster HW, Tierny G, Herman R (1973) Changes in spinal reflex excitability during the initiation of voluntary muscle contraction. Electroenceph Clin Neurophysiol 34: 816–817
11. Paillard J (1955) Réflexe et régulations d'origine proprioceptive chez l'homme. Librairie Arnette, Paris
12. Shimamura M, Akert K (1965) Peripheral nervous relation of propriospinal and spinal-bulbo-spinal reflex system. Jap J Physiol 15: 638–647
13. Shimamura M, Livingstone RB (1963) Longitudinal conduction system serving spinal and brain stem coordination. J Neurophysiol 26: 258–272
14. Sem-Jacobsen CW (1966) Depth electrographic observation related to Parkinson's disease. J Neurosurg 24: 388–402
15. Tóth Sz (1961) Effect of removal of the nucleus dentatus on the parkinsonian syndrome. J Neurol Neurosurg Psychiatry 24: 143–147
16. Tóth Sz (1970) The role of thalamus and pallidum in the activation of the motor system. Confin Neurol 32: 126–127
17. Tóth Sz (1972) Effect of electrical stimulation of subcortical sites on speech and consciousness. In: Soemjen G (ed) Neurophysiology studied on man. Excerpta Medica, Amsterdam, pp 40–46
18. Tóth Sz, Tomka I (1968) Responses of the human thalamus and pallidum to high frequency stimulation. Confin Neurol 29: 17–40
19. Tóth Sz, Zaránd P, Lázár L (1974) The role of the cortex and subcortical ganglia in the evoked rhythmic motor activity. Acta Neurochir (Wien) [Suppl] 21: 25–33
20. Tóth Sz, Zaránd P, Lázár L, Vajda J (1975) Effect of voluntary innervation of the evoked potentials of the motor system. Confin Neurol 37: 49–55
21. Tóth Sz, Vajda J, Zaránd P (1977) The study of recovery modification of the evoked potentials and motor answers of the motor system. Acta Neurochir (Wien) [Suppl] 24: 151–157

22. Tóth Sz, Sólyom A, Vajda J (1979) The frequency resonance investigation of the H-reflex. J Neurol Neurosurg Psychiatry 42: 351–356

23. Tóth Sz, Vajda J, Zaránd P, Sólyom A (1979) Motor regulation in patients with chronic deep electrodes. Rec Dev Neurobiol in Hungary. Academy Publishers, Budapest, pp 73–91

24. Tóth Sz, Vajda J, Sólyom A, Zaránd P (1980) The function of the human cerebellum studied by evoked potentials and motor reactions. Appl Neurophysiol 30: 423–428

25. Tóth Sz, Vajda J (1980) Multitarget technique in Parkinson surgery. Appl Neurophysiol 43: 109–113

26. Tóth Sz, Sólyom A, Vajda J, Tóth Z (1988) The rhythmic properties of the motor system. Appl Neurophysiol in press

27. Upton ARM, McComas AJ, Sica REP (1971) Potentiation of "late" responses evoked in muscles during effort. J Neurol Neurosurg Psychiatry 34: 699–711

Correspondence: Prof. Sz. Tóth, M.D., D.MSc., Neurosurgical Clinic, University of Debrecen, Medical School, Nagyerdei krt. 98, 4012 Debrecen, Hungary.

Acta Neurochirurgica, Suppl. 46, 37–38 (1989)
© by Springer-Verlag 1989

Behavioural Responses to Cerebellar Stimulation in Cerebral Palsy

M. Galanda, L. Mistina, and **O. Zoltan**

Kunz, Banska Bystrica, Czechoslovakia

Summary

Stereotactic placement of stimulating electrodes (TESLA) into deep regions of the cerebellum (Fischer instruments) was analyzed in respect to its organization into sagittaly oriented zones. This was done after imaging of the trajectory and target area and verification of the position of electrodes by CT (Siemens). A correlation was observed between the position of the stimulating electrode from the midline laterally and localization of induced responses on the body—from bilateral to ipsilateral.

The combination of deep cerebellar stimulation and destructive method in the supratentorial region could be the optimal approach to relieve spasticity and to improve motor function in some cases of cerebral palsy.

Keywords: Cerebral palsy; cerebellar stimulation.

Introduction

The determination of the exact target site of stimulation in the cerebellum is crucial for effective therapeutic stimulation and understanding of cerebellar functions in cerebral palsy. Motor responses to electrical stimulation in various subcortical regions of the anterior lobe of the cerebellum in relation to the site of stimulation were investigated. The role of the computerized tomography (CT) scanner is helpful in this respect.

Material and Methods

71 cases of cerebral palsy have been treated with deep cerebellar stimulation via a transentorial approach since 1977[1]. For permanent stimulation system Tesla was used. Trial stimulation at approximately 200 Hz from 0.5–5 mA was always performed and was aimed at chosen targets in the anterior lobe of the cerebellum. The target was determined by reference to the fastigium (F) and the posterior commissure (PC); the motor response to each stimulated site was assessed.

The identification of landmarks was originally done by pneumoencephalography or ventriculography with the patient in the prone position. Today the patient is in the supine position with the head fixed in the Fischer stereotactic apparatus and localization is performed by CT scanner through the reference and target areas in

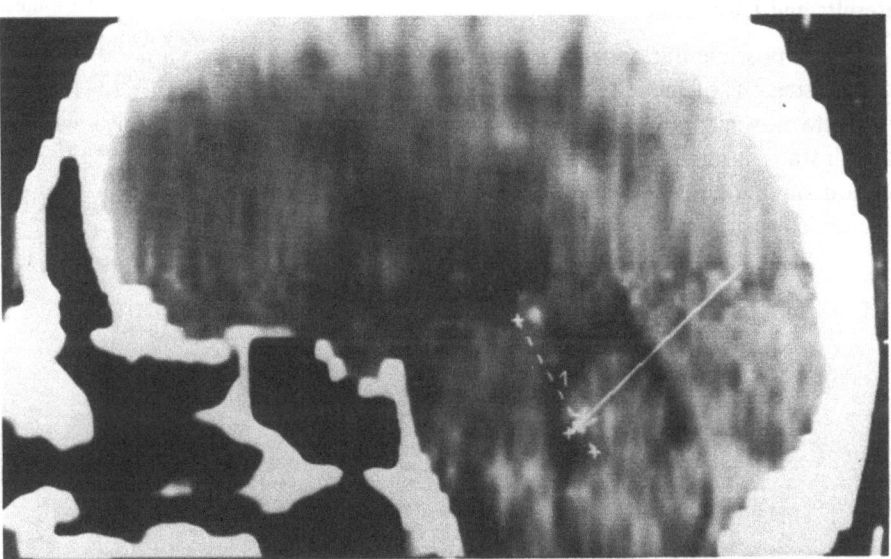

Fig. 1. The CT midline sagittal reconstruction. The trajectory towards target point in the BCC (arrow)

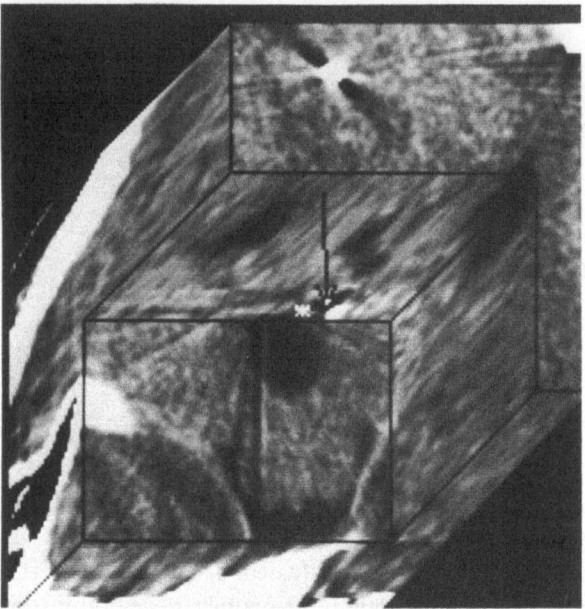

Fig. 2. The CT finding (multiplanar reconstruction) with the electrode in the target point (arrow). Asterisk-the top of the fourth ventricle

2 mm slices. The target is the region of the brachium conjunctivum cerebelli (BCC), 6 mm from the F to the PC, 2 mm below the F-PC line and 4–6 mm lateral to the midline according to the width of the fourth ventricle evaluated at the CT. A sagittal midline reconstruction is performed and F, CP reference points and the target area are calculated (Fig. 1). Laterality is determined on the horizontal slice at the level of the target point identified on the sagittal midline reconstruction. The regions of the cerebellum through which the electrode passes can be shown and the multiplanar programme permits reconstructions in the proposed trajectories of the electrode. The position of the electrode is evaluated by CT examination after its stereotactic insertion and trial stimulation (Fig. 2).

Results and Discussion

It was possible during trial stimulation in the subcortical regions of the anterior lobe of the cerebellum to elicite movement in the extremities, in axial muscles, head rotation, lateral conjugate eye movement, facilitation of involuntary movement and emotional responses (pleasure, fear). The response depended on the

electrode position and the distribution of spasticity and involuntary movement in the body of the patient. Facilitation of muscle tone, beginning in the flexor groups of the extremities, was ipsilateral to the site of the stimulation. If spasticity or involuntary movements were more pronounced contralaterally then the response to stimulation began contralaterally in the more affected site. Rotation of the head and lateral conjugate eye movement always occurred to the ipsilateral side. However, if the stimulating contact was in the midline lateral deviation of the head or eyes did not occur. When the stimulating current was gradually increased (3–5 mA), ophisthotonus often occurred.

During the insertion of the electrode frequent trial stimulation was performed often with correction of electrode position until the proper position was achieved, ignoring previously successful target co-ordinates. Stimulation of BCC produced uniform responses. In our experience if any point in the cerebellum elicits the motor jerk at trial stimulation decrease of pathological hypertonus can be achieved during permanent stimulation[1].

The complicated organization of the cortex and subcortex of the anterior lobe with specific rostrocaudally oriented zones of adjacent but different functions, may be responsible for the unpredictable effect of subcortical stimulation[2]. The proximity of the fourth ventricle to the BCC allows easy identification on the CT and, therefore, the BCC is recommended as an optimal target point for therapeutic stimulation in cerebral palsy.

References

1. Galanda M, Zoltan O (1987) Motor and psychological responses to deep cerebellar stimulation in cerebral palsy. Correlation with organization of cerebellum into sones. Acta Neurochir (Wien) [Suppl] 39
2. Haines DE (1981) Zones in the cerebellar cortex. Their organization and potential relevance to cerebellar stimulation. J Neurosurg 55: 254–264

Correspondence: MUDr. M. Galanda CSc, Kunz, 975 15 Banska Bystrica, Czechoslovakia.

Acta Neurochirurgica, Suppl. 46, 39–45 (1989)
© by Springer-Verlag 1989

Use of Intrathecal Baclofen Administered by Programmable Infusion Pumps in Resistent Spasticity

J. Broseta[*, 1], **F. Morales**[1], **G. García-March**[1], **M. J. Sánchez-Ledesma**[1], **J. Anaya**[1], **J. Gonzalez-Darder**[2], and **J. Barberá**[2]

Departments of Neurosurgery, [1] Hospital Virgen de la Vega, Salamanca, and [2] Hospital Mora, Cadiz, Spain

Summary

On the basis of previous experimental and clinical studies[1, 2] patients with severe spasticity due to spinal cord damage from multiple sclerosis in 8 cases and postraumatic paraplegia in 6 and resistent to all conservative treatments were selected for a trial with morphine and baclofen administered intrathecally through a catheter placed in the spinal subarachnoid space rostral to the affected segments and attached to a subcutaneous reservoir. Whereas morphine single injection did not show any benefit, baclofen bolus injection 30 to 60 µg, revealed a marked decrease of spasticity and associated symptoms in 8 cases. After checking the clinical effect during 3 weeks and changes in electroneurophysiological studies and bladder manometry the catheter was attached to a subcutaneous programmable pump able to be refilled percutaneously and administered baclofen continuously or more often following a multistep complex programme in total doses of 90 to 150 µg per day. After a mean follow-up of 5 months all cases showed an absence of spasms and pain, a notable improvement for bettering of sphincter functions and a marked muscle relaxation that improves motor capacity, leading to increased ambulation or mobility. Neither complications nor side-effects were observed.

Keywords: Spasticity; intrathecal baclofen; infusion pump.

Introduction

Treatment of severe spasticity and related spasms, bladder dysfunction and pain is still a challenge. Oral antispastic drugs have little effect and frequently show side-effects at the required high doses. Neurosurgeons have always been involved in this area and proposed diverse solutions. Ablative procedures and electrical stimulation of the spinal cord have been used to modulate the hyperactive response of spasticity but with inconsistent results.

Baclofen, a gamma aminobutyric acid analogue, was introduced in the seventy to manage severe spasticity, showed successful results as an oral agent[7] but with frequent side-effects and limitations due to the high plasma levels to cross to blood brain barrier[3, 4]. When Kroin et al.[5] observed that intrathecal baclofen caused a significant reduction of the polysynaptic reflexes in rabbits a new alternative to modulate spasticity emerged. Penn and Kroin[9] introduced spinal intrathecal baclofen for severe spastic conditions and reported excellent initial results proposing the use of programmable devices for chronic infusion. This promising early experience was later confirmed by other groups although with overdose problems[8, 10]. Other drugs such as morphine and benzodiazepines have been also infused intrathecally in an attempt to control spasticity but showed irregular results[2, 6, 10].

We present the results in 12 cases with marked spinal spasticity treated with spinal intrathecal baclofen given in a single bolus injection and chronic infusion by programmable pumps.

Clinical Material and Methods

Patients: Twelve patients with severe spasticity unresponsive to oral medication or with side-effects were selected for trial. Table 1 summarizes the clinical material and response of this group of 6 females and 6 males with ages from 19 to 61 years. The cause of spasticity was multiple sclerosis in 5 cases, cord trauma in 5, cervical spondyloarthrotic myelopathy in 1 and syringomyelia in 1. The duration of the illness varied from 10 months to 23 years. The poor neurological status contributed to poor physical performance, transfers, daily activities and sleep. There was paraparesis in 7 cases, paraplegia in 4 and tetraplegia in 1 so that 6 cases were reduced to a wheelchair or bedridden state. The remainder had partial mobility using braces or canes. The majority of patients showed high scores

* J. Broseta, M.D., Cátedra de Neurocirugía, Facultad de Medicina, c/Espejo, E-37007 Salamanca, Spain

Table 1 a. *Spasticity History, Neurological Situation and Response to Trial with Intrathecal Drugs*

Case	A/S	Cause	Duration	Spasticity status		
				Condition	Rigidity	Spasms
1	26/F	Th_3 - Th_4 cord trauma	5 years	Paraplegia. Non mobile.Clonus Hyperreflexia. Wheelchair	4	>2 per h.
2	52/F	Multiple sclerosis	7 years	Paraparesis. Mobile in braces. Mild hyperreflexia	3	>1 per h.
3	49/M	Th_{11} - L_1 cord trauma	3 years	Paraparesis. Mobile in braces with orthesis. Mild hyperreflexia	3	<1 per h.
4	51/F	Multiple sclerosis	19 years	Paraparesis. Non mobile.Clonus Hyperreflexia. Wheelchair	3	Occasional
5	32/M	Th_{12} - L_1 cord trauma	10 months	Paraplegia. Non mobile.Clonus Hyperreflexia. Wheelchair	4	Occasional
6	58/M	Cervical myelopathy	7 years	Mild paraparesis. Hyperreflexia Clonus. Handicaped walking.	2	Occasional
7	19/M	C_6 cord trauma	11 months	Tetraplegia. Bedridden	4	Occasional
8	43/M	Multiple sclerosis	10 years	Paraparesis. Mobile in canes. Hyperreflexia	4	Occasional
9	61/F	Th_5 - Th_6 cord trauma	9 years	Paraplegia. Non mobile.Clonus Hyperreflexia. Wheelchair	5	>1 per h.
10	27/F	Multiple sclerosis	4 years	Mild paraparesis. Partial mobility. Mild hyperreflexia. Handicaped walking	2	>1 per h.
11	33/M	Syringomyelia	11 months	Mild paraparesis. Partial mobility. Mild hyperreflexia. Handicaped walking	2	no
12	43/F	Multiple sclerosis	23 years	Paraplegia. Non mobile.Mild hyperreflexia. Bedridden	3	<1 per h.

in the Ashworth scale[3] for rigidity and all but one had disturbing spasms. Neurological bladder dysfunction was present in 9 cases with incontinence and alterations of the micturation. Cramping or diffuse pain in low back and/or lower extremities was present in 7 cases and speech impairment in 2. In some cases bladder manometry and perineal electromyographic recordings were performed to evaluate changes after intrathecal baclofen. In 5 cases the Hmax:Mmax ratio was also determined for further comparison.

Trial period: This group had a preliminary trial of intrathecal antispastic drugs via a subcutaneous port attached to a catheter introduced in the lumbar subarachnoid space and manipulated to the affected cord segments. Intrathecal bolus of morphine and baclofen were tested through this system for several consecutive days observing the effect on spasticity and associated symptoms. In the first cases morphine offered uncertain results. Baclofen, however,

always had alleviation of spastic symptoms. Baclofen* was diluted in saline to a concentration of 50 µg per ml. Trial started with a bolus of 12.5 µg, with increases of dose every second day. Reduction of rigidity and spasms, pain remission and improvement of bladder function occurred for 7 hours after the bolus in 8 out of 12 patients. Weakness, drowsiness, mental confusion, respiratory depression or lethargy did not appear. These 8 cases were considered as suitable candidates for chronic infusion and fulfilled the criteria of severe spasticity resistent to oral medication, normal anatomy of the spinal canal and CSF circulation and a positive response to a dose less than 100 µg.

---

* Baclofen was received by courtesy of Ciba-Geigy, Barcelona, Spain.

Table 1 b. *Spasticity History, Neurological Situation and Response to Trial with Intrathecal Drugs*

| Case | Neurogenic bladder dysfunction | Other symptoms | Drug effect during trial | | Programmable infusion pump |
			Morphine	Baclofen	
1	yes	Pain	Partial improvement	Improvement	Implanted
2	yes	Pain	No change *	Improvement	Implanted
3	Sphincterotomy	Pain	Partial improvement	Improvement	Implanted
4	yes	Speech impairment	No change	No change	Not implanted
5	yes	Pain	No change*	Improvement	Implanted
6	no			Improvement	Implanted
7	yes		No change	No change	Not implanted
8	yes		No change	Improvement	Implanted
9	yes	Pain		No change	Not implanted
10	no	Pain		Improvement	Implanted
11	no	Pain		No change	Not implanted
12	yes	Pain. Lupus. Speech impairment. Upper limb affection.		Improvement	Implanted

* no change but pain.

Chronic infusion: Programmable infusion pump** implantation was done under sedation and local anaesthesia. After preparing the device in the proper conditions, the liquid contents was withdrawn and replaced by 18 ml of 500 µg per ml baclofen. Meanwhile purging, subarachnoid catheter was isolated, the previous subcutaneous port removed and the pump attached. The device was placed and anchored in a deep subcutaneous abdominal pocket in to allow later percutaneous refilling. At the conclusion of the operation several parameters concerning general data, specific properties of the drug, flow conditions and alarm activation were transmitted from an external programmer to the implanted pump.

Results

Table 2 illustrates the results. The spastic symptoms generally improved in the majority of cases. A reduction of at least one score in the Ashworth scale was a

** Programmable infusion pumps were provided by Medtronic Inc., Minneapolis, Minn.

long-lasting finding in the 8 cases but less than that during the trial (Fig. 1 a). Spasms were also drastically alleviated. Four patients that previously showed an average of more than one spasm per hour are free of them following treatment (Fig. 1 b). This improvement also involved the hyperreflexive conditions that changed to a mild hyperreflexive state (Fig. 1 c). Despite neurological and subjective amelioration this was not always associated with better motor improvement. Thus, only the 4 less affected cases increased the walking ability, transfers and daily activities. The rest gained in life quality and comfort (Fig. 1 d).

In 2 cases electroneurophysiologic studies following chronic baclofen infusion showed a significant reduction of 23 and 50% respectively in Hmax:Mmax ratio. In the other 3 cases these changes were less marked. All cases had long-lasting improvement in bladder

Table 2. *Results After Chronic Intrathecal Infusion of Baclofen Using Programmable Devices*

Case	Programmed infusion parameter			Result
	Total dose µg. per day	Mode	Partial dose	
1	60 µg.	Continuous	2.5 µg./ hour	Reduction in rigidity from 4 to 2. No spasms. Pain remision.Improvement in bladder control. Better performance in physiotherapy
2	96 µg.	Continuous	4 µg./ hour	Moderate reduction in rigidity from 3 to 2. No spasms. No pain. Better bladder control. Increase in walking capacity.
3	120 µg.	Continuous	5 µg./ hour	Reduction in rigidity from 3 to 1. No spasms. Occasional pain. Much better performance in physical activities.
5	80 µg.	Bolus	20 µg./ 6 hours	Moderate reduction in rigidity from 4 to 3. No spasms. No pain. Better bladder control
6	96 µg.	Continuous	4 µg./ hour	Moderate reduction in rigidity from 2 to 1. No spasms. Increase in walking capacity
8	100 µg.	Bolus	25 µg./ 6 hours	Moderate reduction in rigidity from 4 to 3. No spasms. Improvement in bladder control. Increase in walking capacity
10	148 µg.	Multistep complex	2 µg./ hour 25 µg./ 6 hours	Reduction in rigidity from 2 to 1. Occasional spasms. No pain. Almost normal walking
12	172 µg.	Multistep complex	3 µg./ hour 25 µg./ 6 hours	Moderate reduction in rigidity from 3 to 2. Occasional spasms. No pain. Better bladder control. Better speech and performance with upper limbs

function expressed in a regular micturition rhythm and decrease of incontinence subsequently resulting in less infection. Bladder manometry and perineal electromyographic recordings showed normal values in basal pressure, a significant increase in bladder capacity and decrease in contraction and less abnormal discharges following an intrathecal bolus of 20 µg baclofen (Fig. 2).

To obtain these promising results the total dose per day of baclofen was individually estimated from the effective dose during the trial and according to the mode of infusion. The initial total dose varied from 60 to 172 µg. But a progressive increase of dose and change of the mode was required to maintain the initial efficacy, probably due to tolerance (Fig. 3). In this period, neither malfunction nor drug overdose occurred. Complications were only a pocket seroma in one case and skin erosion on the pump attachment in another, both probably related to the still imperfect technological design of the device regarding size, anatomical accommodation, reservoir capacity and location of the metalic apendix for drainage.

Discussion

In our hands chronic intrathecal infusion of baclofen using programmable pumping devices has been a promising approach to reduction of rigidity, hyperreflexive status, spasms, bladder dysfunction and to slightly improve motor function and mobility. Despite these encouraging results there are still several problems to be solved: to devise standard protocol for selection of patients, to study the appropriate dose and

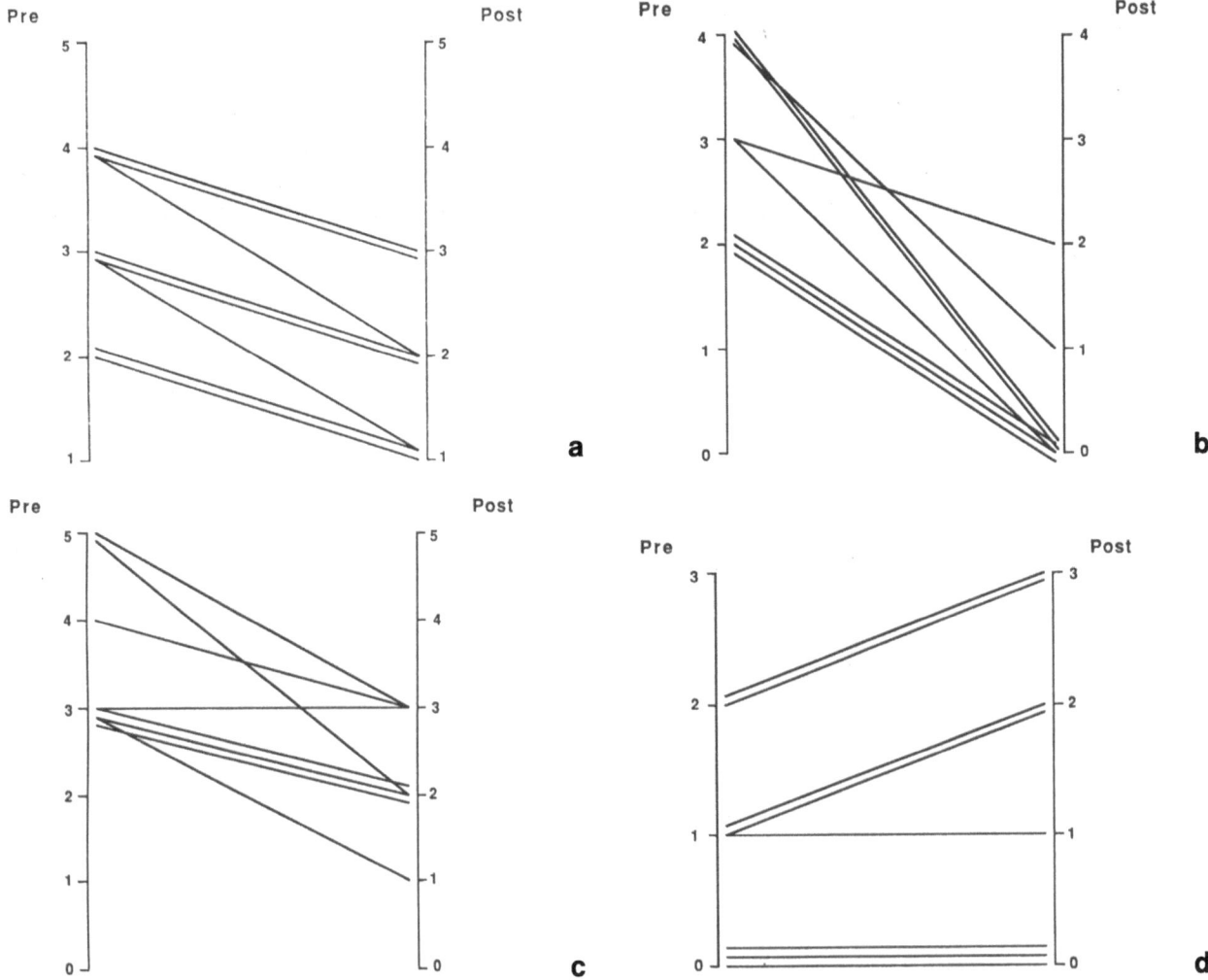

Fig. 1. Results on spastic and related symptoms after following intrathecal infusion of baclofen for a mean period of 5 months. a) Changes in rigidity scored by the Ashworth Scale: *1* no increase in tone; *2* slight increase in tone; *3* marked increase in tone but limbs easily flexed; *4* passive movement difficult; *5* rigidity in flexion and extension. b) Changes in spasm occurrence: *0* none; *1* spasms after motor and sensory stimulation; *2* occasional spontaneous spasms; *3* less than 1 spontaneous spasm per hour; *4* more than 1 spontaneous spasm per hour. c) Changes in reflexes: *0* absent; *1* hyporreflexive; *2* normorreflexive; *3* mild hyperreflexive; *4* hyperreflexive; *5* clonus. d) changes in voluntary mobility and physical performance: *0* non mobile, weelchair or bedridden, dependent of assistance for skilled acts; *1* partial mobility, braces or canes, partial dependent of assistance for skilled acts; *2* handicaped walking and independent for skilled acts; and *3* normal walking and independent for normal activities

mode of infusion to maintain the efficacy and avoid tolerance and to improve the design of the pumps.

The criteria for selection of candidates were initially limited to cases with severe spasticity of spinal cord origin resistent to oral medication. Nevertheless improvement of speech impairment occurred in 2 cases with multiple sclerosis and of motor performance of the upper limbs in, 1 with multiple sclerosis and 1 with postraumatic tetraplegia, even though baclofen was de-

livered caudally to the affected segment. Dralle *et al.*[1] also found that lumbar intrathecal baclofen alleviated spastic conditions caused by diffuse cerebral damage in children and adults.

In our experience during the trial period as well as during chronic infusion special attention was given to the estimation of the effective dose, since drug overdose is still an important complication when a suitable antagonist is not yet available.

Bladder manometry

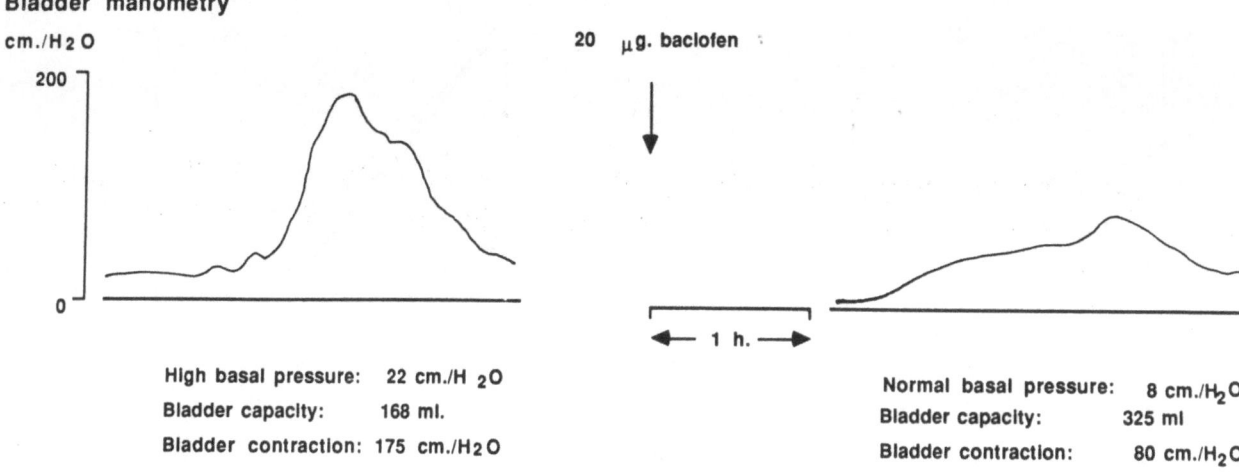

High basal pressure: 22 cm./H$_2$O	Normal basal pressure: 8 cm./H$_2$O
Bladder capacity: 168 ml.	Bladder capacity: 325 ml
Bladder contraction: 175 cm./H$_2$O	Bladder contraction: 80 cm./H$_2$O

Perineal electromyographic recording

Fig. 2. Changes in bladder function following an intrathecal bolus of 20 µg baclofen

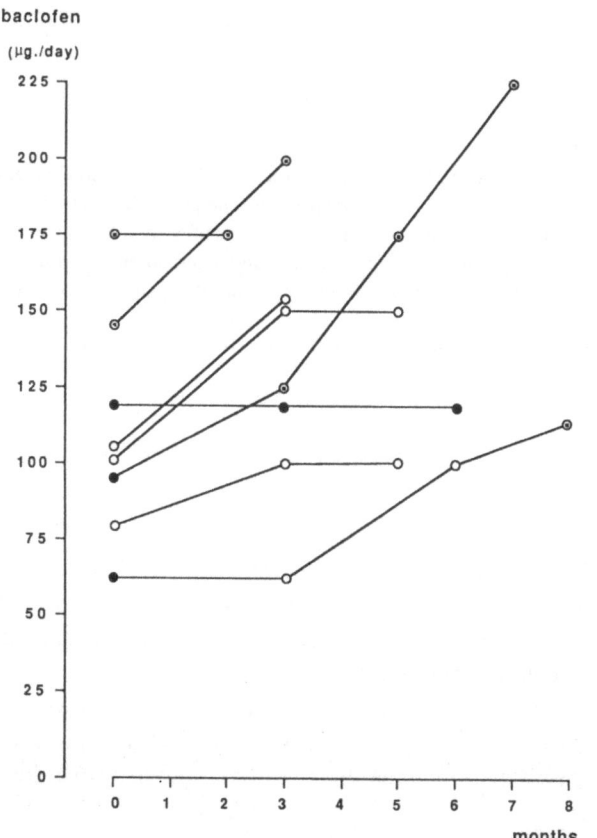

Fig. 3. Intrathecal baclofen dosages and mode of infusion to maintain the initial outcome: ● continuous mode; ○ bolus; ⊙ multistep complex mode

Acknowledgements

In this study we acknowledge Paco Romero for introducing us in the complex technique of programming the infusion pumps. We also are indebted to Francisco Sabidó for the facilities given in obtaining baclofen.

References

1. Dralle D, Muller H, Zierski J *et al* (1985) Intrathecal baclofen for spasticity. Lancet 2: 1003
2. Erickson DL, Blacklock JB, Michaelson M *et al* (1985) Control of spasticity by implantable continuous flow morphine pump. Neurosurgery 16: 215–217
3. Hattab JR (1980) Review of european clinical trials with baclofen. In: Feldman RG, Young RR, Koella WP (eds) Spasticity: disordered motor control. Miami: Symposia Specialists, pp 71–75
4. Knuttson E, Lindblom U, Martenson A (1974) Plasma and cerebrospinal fluid levels of baclofen (Lioresal) at optimal therapeutic responses in spastic paresis. J Neurol Sci 23: 473–484
5. Kroin JS, Penn RD, Beissinger RL *et al* (1984) Reduced spinal reflexes following intrathecal baclofen in the rabbit. Exp Brain Res 54: 191–194
6. Muller H, Gerlach H, Boldt HJ (1983) Aufhebung spinaler Spastik durch intrathekale Benzodiazepin-Applikation. Anaesthesist [Suppl. ZAK 83] 32: 17–19
7. Pederson S, Arlien-Soborg P, Mai J (1974) The mode of action of the GABA derivative baclofen in human spasticity. Acta Neurol Scand 50: 665–680
8. Penn RD, Kroin JS (1987) Long term intrathecal baclofen infusion for treatment of spasticity. J Neurosurg 66: 181–185
9. Penn RD, Kroin JS (1985) Continuous intrathecal baclofen for severe spasticity. Lancet 2: 125–127
10. Siegfried J, Lazorthes Y (1985) Neurochirurgie de l'infirmité motrice cérébrale. Neurochirurgie 31: 95–101
11. Siegfried J, Rega GL (1987) Intrathecal application of baclofen in the treatment of spasticity. Acta Neurochir (Wien) [Suppl] 39: 121–123

Correspondence: J. Broseta, M.D., Cátedra de Neurocirugía, Facultad de Medicina, c/Espejo, E-37007 Salamanca, Spain.

Acta Neurochirurgica, Suppl. 46, 46–47 (1989)
© by Springer-Verlag 1989

Critical Approach to Intrastriatal Medullary Adrenal Implants via Open Surgery in Parkinsonism. Case Report

J. Broseta*, P. Diaz-Cascajo, G. García-March, and **M. J. Sánchez-Ledesma**

Departments of Neurosurgery, Hospital Virgen de la Vega, University of Salamanca, Salamanca, Spain

Summary

Encouraged by the recent reports on the beneficial effects obtained with open transplantation of autologous adrenal medullary grafts into striatal structures in cases with resistent Parkinson's disease, our team used this procedure in a 63-year-old man presenting with severe bradykinesia and rigidity resistent to all pharmacological attempts. In this case through a laparotomy the right adrenal gland was removed and stored in oxygenated Collins and bicarbonated Ringer solution mixtures while a F_2 transventricular approach to the head of the caudatum was done. With the surgical microscope the medullary part of the adrenal gland was dissected and four pieces of mm^3 of tissue selected, implanting them in a bed previously carved in the caudatum. Endocrinologic and hydroelectrolytic problems appeared during the immediate postoperative period. In the following 5 months no clinical benefit nor electroneurophysiological changes were observed.

Keywords: Medullary adrenal grafts; striatum; Parkinson's disease.

Introduction

In rats with 6-hydroxydopamine nigrostriatal denervation rotational behaviour returns to normal after fetal substantia nigra[2, 4, 5] or adrenal medullary tissue[6, 9] is transplanted to the caudatum. Backlund *et al.*[1] introduced intrastriatal medullary adrenal grafting for patients with resistent parkinsonism. Several years later, Madrazo *et al.*[8] revised this procedure and reported significant early improvements in two young patients with intractable Parkinson's disease. They incorpored some technical variations, namely the use of open surgery to approach the head of the caudatum and the placement of the grafted tissue in contact with the cerebrospinal fluid since there was a better growth and survival of the transplants[7, 10, 11].

* J. Broseta, M.D., Cátedra de Neurocirugía, Facultad de Medicina, c/Espejo, E-37007 Salamanca, Spain.

Encouraged by these reports a patient was selected for the application of this surgical variation, despite Backlund's unsatisfactory experience and our poor results in an experimental study showing the low survival rate of medullary adrenal tissue implanted in the striatum[3].

Case Report

The patient was a 61-year-old retired farmer with 4 years history of Parkinson's disease, who had rapidly deteriorated in the previous 7 months, resistent to oral medication of 1,000 mg. L-dopa with decarboxylase inhibitor, 60 mg of bromocriptine and the usual dose of anticholinergics which caused severe side-effects. He had no history of drug abuse or familial parkinsonism. When admitted to the hospital he was extremely disabled with generalized rigidity, bilateral resting tremor predominantly in left extremities, severe akinesia with masked facial expression, speech impairment and global incapacitation for skilled acts which made him completely dependent and confined to a wheelchair.

Before surgery was decided, a psychological evaluation excluded dementia. The adrenal function as well as brain and adrenal CT scannings were normal. Electromyographic recordings showed fine tremor of 3 to 5 cps in both flexor and extensor musculature of the left arm. Homovanillic (31 µg/ml) and 5-hydroxy-indolacetic (0.035 µg/ml) acids were estimated in lumbar CSF to assess monoamine metabolism. After surgery was recommended, the pro and cons of the operation were explained to the patient and relatives and written consent obtained.

The surgical technique described by Madrazo *et al.*[8] was followed with minor variations. A right adrenalectomy was done through an anterior laparotomy and a right frontal craniotomy for frontal lobe exposure were performed simultaneously. After removal, the gland was stored in oxygenated bicarbonate glucose Ringer and Collins solutions mixed at 50% whilst waiting for microdissection. Meanwhile, with the surgical microscope the head of the right caudatum was approached through the lateral ventricle by means of a F_2 transcortical dissection. Once the adrenal gland was cut sagitally, the medullar was identified and dissected from the cortex, cut in several fragments of approximately 1 mm^3 with the aid of the microscope and placed on a wet surface containing the culture solution. The

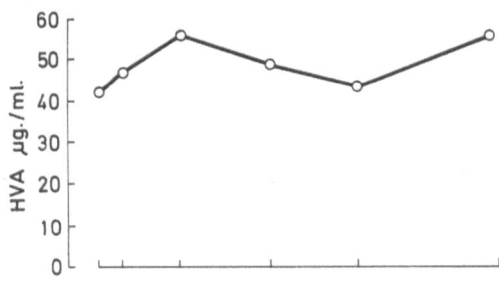

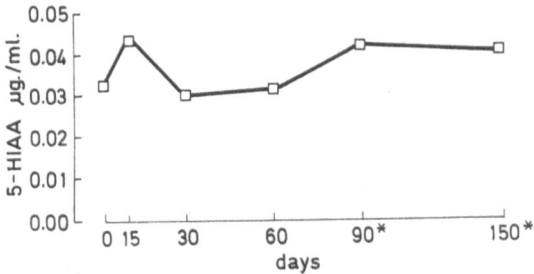

Fig. 1. Postoperative changes in concentration of homovanillic (HVA) and indolacetic (5-HIAA) acids in lumbar cerebrospinal fluid. * Pharmacological treatment reinstated

abdominal surgeons replaced the rest of the gland in the abdominal wall, between the muscular and subcutaneous layers, in an attempt to preserve cortical adrenal function. Simultaneously, the neurosurgical team made a small bed in the head of the caudatum and four medullary adrenal fragments were transplanted there. To avoid migration the bed was sealed with a small piece of oxydized cellulose net anchored to the parenchyma with hemoclips, in this way keeping the grafted tissue in contact with the CSF.

After a stormy postoperative course of 5 months, the patient died of pulmonary embolism. During this time there was no neurological, neurophysiological nor subjective improvement. For that reason, treatment with oral L-Dopa and bromocriptine was reinstated 3 months postoperatively. Postoperative brain CT scans showed no evidence of the grafted tissue but a hypodense image occupied the head of the right caudatum which could be interpreted as necrosis. In order to know whether the grafted tissue was functionally active or not, CSF concentrations of homovanillic and indolacetic acids were periodically determined without finding significant changes when compared with the previous values (Fig. 1).

The patient remained hospitalized for 5 months, 3 of them in the Intensive Care Unit. He awoke promptly after surgery and the immediate postoperative period was uneventful, though he developed adrenal insufficiency with electrolyte disorders 24 hours later. During this time, the patient was totally conscious and responsive. In the next week, a pneumonia and femoral thrombophlebitis appeared which led progressively to serious respiratory problems. These continued over the following 2 months with a candida pneumonia, atelectasis and pulmonary embolism due to an axillary thrombophlebitis, needing mechanical ventilation for several days. Five months postoperatively he had apparently overcome this poor status when he suddenly died of a second massive pulmonary embolism.

Remarks

Neural or chromaffin tissue implants into the brain are a promising alternative for treating neurological disorders but in accordance to the controversial reported results it has probably been too promptly incorporated to the human clinic. Several points ought be considered: a neural grafting general policy as other specialists did for other organ transplants; the use of homologous fetal or adult chromaffin or nervous tissue grafts after consideration of the immunological and ethical issues; the proper target for implantion; the use of stereotactic versus open surgery; and, finally to devise a protocol for indications and selection of candidates.

References

1. Backlund EO, Granberg PO, Hamberger B *et al* (1985) Transplantation of adrenal medullary tissue to striatum in parkinsonism. J Neurosurg 62: 169–173
2. Björklund A, Dunnett SB, Stenevi U *et al* (1980) Reinnervation of the denervated striatum by substantia nigra transplants: functional consequences as revealed by pharmacological and sensorimotor testing. Brain Res 199: 307–333
3. Broseta J, Diaz-Cascajo P, García-March G *et al* Intrastriatal adrenal medullary grafts in rats with 6-hydroxydopamine damaged nigrostriatal system: graft survival rate and changes in rotational behaviour. Appl Neurophysiol (in press)
4. Dunnett SB, Björklund A, Stenevi U *et al* (1981) Behavioural recovery following transplantation of substantia nigra in rats subjected to 6-OHDA lesions of the nigrostriatal pathway. I. Unilateral lesions. Brain Res 215: 147–161
5. Freed WJ, Perlow MJ, Karoum F *et al* (1980) Restoration of dopaminergic function by grafting of fetal rat substantia nigra to the caudate nucleus: long-term behavioral, biochemical, and histochemical studies. Ann Neurol 8: 510–519
6. Freed WJ, Morihisa JM, Spoor E *et al* (1981) Transplanted adrenal chromaffin cells in rat brain reduce lesion-induced rotational behaviour. Nature 192: 351–352
7. Freed WJ, Cannon-Spoor HE, Krauthamer E (1986) Intrastriatal adrenal medullar grafts in rats: long-term survival and behavioral effects. J Neurosurg 65: 664–670
8. Madrazo I, Drucker-Colin R, Diaz V *et al* (1987) Open microsurgical autograft of adrenal medulla to the right caudate nucleus in two patients with intractable Parkinson's disease. New Engl J Med 316: 831–834
9. Olson L, Hamberg B, Hoffer B *et al* (1981) Nerve fiber formation by grafted adult adrenal medullary cells. In: Stjärne J, Hedqvist P, Lagercrantz H *et al* (eds) Chemical neurotransmission 75 years, Second Nobel Conference. Academic Press, London, pp 35–48
10. Perlow MJ, Kumakura K, Guidotti A (1980) Prolonged survival of bovine adrenal chromaffin cells in rat cerebral ventricles. Proc Natl Acad Sci USA 77: 5278–5281
11. Rosestein JM, Brightman MW (1978) Intact cerebral ventricle as a site for tissue transplantation. Nature 276: 83–85

Correspondence: J. Broseta, M.D., Cátedra de Neurocirugía, Facultad de Medicina, c/Espejo, E-37007 Salamanca, Spain.

Acta Neurochirurgica, Suppl. 46, 48–50 (1989)
© by Springer-Verlag 1989

Transplantation in Parkinson's Disease: Stereotactic Implantation of Adrenal Medulla and Foetal Mesencephalon

E. R. Hitchcock, C. G. Clough, R. C. Hughes, and **B. G. Kenny**

Department of Neurosurgery, University of Birmingham, Midland Centre for Neurosurgery and Neurology, Birmingham, U.K.

Summary

The two possible dopamine donor sites for transplantation are autologous adrenal medulla and human foetal substantia nigra or adrenal medulla. There is increasing experience with the use of adrenal medulla transplantation for Parkinson's disease and much less experience in foetal substantia nigra transplantation.

The particular problems of each technique are discussed with examples and postoperative progress of cases with the special emphasis on management problems.

Keywords: Parkinson's disease; STIM; transplant; foetal.

Laboratory work has shown that brain tissue grafts develop connections to the host brain and can produce behavioural changes. The nigrostriatal system is a particularly suitable model to demonstrate the influence of neural grafting on disorders induced by particular central nervous system lesions[2, 9]. Adrenal medullary grafts into the striatum do not innervate the host although they may abolish rotational behaviour produced by substantia nigral lesions, by diffusion of dopamine into the striatum. Foetal substantia nigra transplants do innervate the implantation site and these experiments have encouraged their application in dopamine deficient disorders such as Parkinson's disease.

In 1985 two patients with severe Parkinson's disease were reported to have had autologous adrenal medullary tissue transplanted unilaterally to the putamen with brief improvement[1, 5]. In 1987 autologous adrenal medulla was transplanted into the head of the caudate nucleus via craniotomy[6] and its remarkable success was followed by further grafts with good results in young patients but a high mortality rate and morbidity in elderly patients. The alternative procedure of foetal tissue transplantation entails less surgical risk and the same surgeon treated two patients with severe Parkinson's disease with foetal substantia nigra and adrenal medulla transplants respectively[7]. Both patients had improved eight weeks after the procedure.

We report our early and limited experience with both methods of transplantation. Application for the adrenal transplantation was made to a duly constituted hospital ethical committee in May 1982 and permission granted for this procedure in February 1987. Application was made in October 1986 for the foetal transplantation which was granted in December 1987.

Autologous Adrenal Medullary Transplantation (AAMT)

Case History: A 45-year-old man presented with right sided tremor due to Parkinson's disease six years previously. Relatively well controlled on increasing doses of Madopar (L-dopa and Benserazide) 125 mg six times daily, he had a residual right arm tremor with mild rigidity and a degree of bradykinesia (Webster Rating Scale-WRS 10/30) but he was fully independant (North-Western University Disability Score, NUDS = 0/50). Because of progressive deterioration he presented himself for consideration of implantation. Two months prior to admission an attempt made to stop his Madopar therapy failed because of increased right sided tremor and incapacity.

Method: On 12th October 1987 CT scan of the abdomen confirmed the presence of bilateral adrenals. The stereotactic square was fixed to the patients head under local anaesthesia and stereotactic CT targeting of the head of the right caudate nucleus targeted by CT stereotaxy. The patient was anaesthetized and the left adrenal exposed and removed by laparotomy. The adrenal medullary tissue was implanted stereotactically into the head of the right caudate nucleus as a solid mass, part of which projected into the ventricle and secured to the ependymal edge by a silver clip. A second medullary fragment was secured to the choroid plexus. Postoperative recovery was swift and uncomplicated.

Progress: Immediately post-operatively, without medication, the patient had marked improvement in tremor and bradykinesia medication but within 36 hours his Parkinsonism had deteriorated with

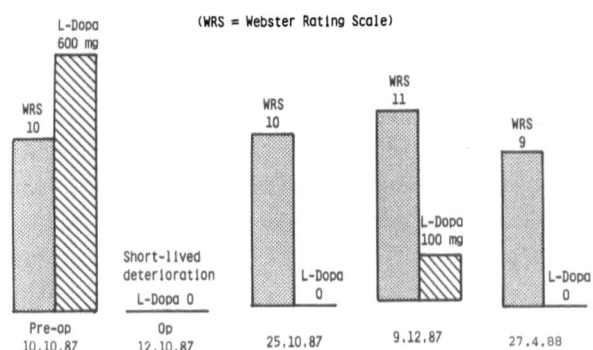

Fig. 1. J. D.—implantation of adrenal medulla

increased tremor and bradykinesia. After one week without medication his Parkinsonian ratings had returned to preoperative rating with medication. Two months later he remained well although he had re-started Madopar 125 at night because of "stiffness" in the right side which he stopped in January and on the 27th April 1988 WRS = 9/30 and the same as preoperative ratings on L-dopa (Fig. 1).

Stereotactic Implantation of Foetal Mesencephalon (STIM)

Dead aborted foetuses collected routinely for other research purposes were examined over an 18-month period to establish a routine for the identification and isolation of the foetal mesencephalon. Research was conducted strictly according to the code of practice outlined in the Peel report (1972). The mesencephalic tissue was kept in a tissue medium with antibiotic and disaggregated mechanically into a thick cell suspension shortly before injection.

Case Report 1: A 60-year-old woman with Parkinson's disease for 25 years had gradual deterioration with a dramatic on/off syndrome over the last 12 years as a response to her dopaminergic therapy. During an "off" phase she was immobile and unable to move from a chair (Hoehn and Yahr grade V). With one assistant she could walk with six inch steps but would promptly fall if left unattended. During an "on" phase she had chaotic chorea of all four limbs and neck but functional improvement enabled her to walk unattended (Hoehn and Yahr grade IV) although still very unsteady with many spectacular falls. During an "on" phase she could dress herself with assistance and eat slowly although her voice never became more than a hurried almost inaudible whisper. Many different drug treatments had been carried out without great improvement. She remained highly sensitive to L-dopa preparations requiring small, frequent doses but reached an optimum dose of Madopar 62.5 taken every one and a half hours. Attempts at reducing L-dopa therapy proved disasterous and life threatening. Preoperatively she took Madopar 62.5 × 14 capsules daily and Bromocriptine 5 mg four times daily. Preoperative WRS averaged $^{29}/_{30}$ during "off" phase and $^{20}/_{30}$ during "on" phase. NUDS $^{36}/_{50}$ during "off" phase, $^{20}/_{50}$ during "on" phase. "On" hours averaged 8–12 per day.

Method: On March 7th the foetal suspension was implanted stereotactically into the head of the right caudate nucleus via a right

frontal burr hole after stereotactic CT scanning. This was performed under local anaesthetic without postoperative complications. A similar method was used for Case 2 on 7th April 1988.

Progress: Improvement appeared within hours and all anti-Parkinsonian medication was stopped four days later. The patient remained profoundly Parkinsonian (WRS $^{29}/_{30}$) but was able to walk without assistance (Hoehn and Yahr grade IV), feed herself slowly and carry out all activities of daily life. Previous preoperative attempts to withdraw L-dopa treatment had resulted in catastrophic life-threatening akinesia. In view of continued Parkinsonism Madopar 62.5 three times daily was introduced on the 6th April which produced improvement but the onset of right leg dyskinesia; increasing dosage merely increased her dyskinesia without improving function. Assessment of Parkinsonism showed WRS $^{20}/_{30}$ and NUDS $^{21}/_{50}$. Using this small dosage of L-dopa there is little discernible fluctuation and only minimal dyskinesia as above. Function improved, particularly speech (Dysarthria Assessment Scale), and the patient reported great benefit. However, other objective rating scales revealed no difference between scores postoperatively and those of the pre-operative "on" state. The rewards are the absence of profound Parkinsonian "off" state and a continuous uniform condition throughout the day without chaotic dyskinesia (Fig. 2).

On 26th May 1988 she reported that she had reduced her Madopar to two tablets daily because she remained active but had developed "off" symptoms such as she had experienced previously on high L-dopa reduction.

Case Report 2: A 41-year-old man developed Parkinson's disease at 35 starting with right hand tremor. There was a good initial response to L-dopa therapy but recent increasing deterioration required increased doses of L-dopa. Because of his poor prognosis he presented himself for implantation therapy. He remained fully functional and ran his own business and during his leisure hours could act in Gilbert and Sullivan operatic productions. Drug therapy preoperatively was Madopar 250 4-hourly, Benzhexol 2 mg four times a day. Each Madopar exerted it's best effect over 3–4 hours. One hour post-dose WRS $^9/_{30}$, NUDS $^3/_{50}$, Hoehn and Yahr grade III. He was taken off all treatment two weeks pre-operatively and became profoundly Parkinsonian needing help to get out of a chair and help

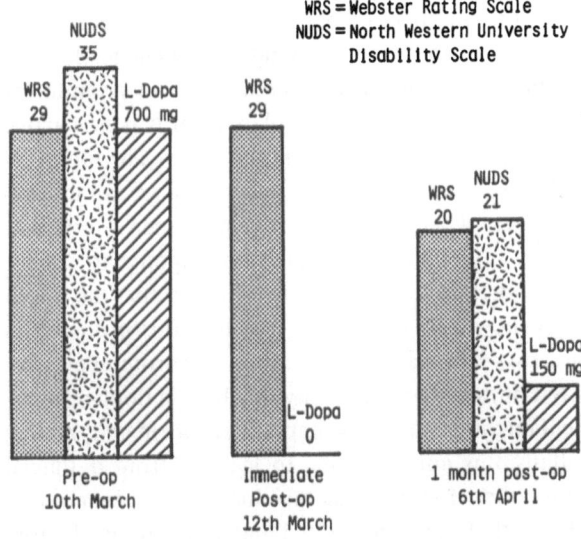

Fig. 2. E. K.—stereotactic implant of fetal mesencephalon

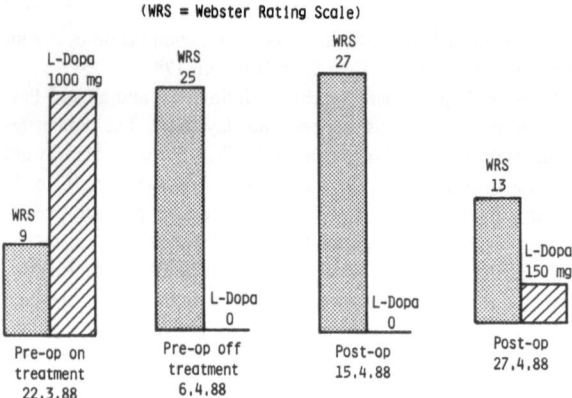

(WRS = Webster Rating Scale)

Fig. 3. R. D.—stereotactic implant of fetal mesencephalon

with all activities of daily life, WRS $^{25}/_{30}$, NUDS $^{31}/_{50}$, Hoehn and Yahr grade V. The implantation into the right head of caudate took place on the 7th April.

Progress: Post-operatively he remained well with improvement in his akinesia, however WRS remained essentially unchanged, $^{26-}$ $^{27/30}$. Nine days post-operatively re-introduction of Madopar 62.5 tds produced no improvement over 24 hours. Benzhexol 2 mg tds (Cholinergic) produced a dramatic improvement within 8 hours suggesting an anti-cholinergic withdrawal phenomenon. WRS on 22nd April 1988 and 27th April 1988 being $^{13}/_{30}$ with return to previous full function, NUDS $^{4}/_{50}$50, Hoehn and Yahr grade III (Fig. 3). On the 26th May he reported that he remained active on one sixth of his previous L-dopa dosage. WRS = $^{11}/_{30}$, NUDS $^{5}/_{50}$, Hoehn and Yahr grade III.

Discussion

Our experience with both types of procedure were communicated to the Society of British Neurological Surgeons meeting in Oxford in April 1988[8]. We would emphasize that the treatment is entirely experimental and large numbers of patients are needed before the techniques can be recommended as beneficial to all patients with Parkinson's disease. These preliminary results demonstrate that implantation can be carried out safely without major complications. The initial results have been encouraging but as yet we cannot definitely say whether any improvement seen is attributable to the implant. The placebo response to such a brain operation could be powerful and may be implicated, however, we believe that had this occurred in case two it would have become less prominent after two months. More hopefully the drop in dopa requirement might be due to the implant producing dopamine; initially this could be due to leakage from damaged cells rather than intrinsic production. Another possibility is that the transplanted material contains factors which activate the recipient striatal neurones. A prob-

lem in interpretation of the lower dopa requirements is that all three patients have undergone a dopa holiday and diminished requirements following such a holiday (perhaps due to alterations in the dopamine receptor) are well recognized[4, 8]. Time will tell whether this drop in dopa requirements (to one fifth in both patients with STIM and to no medication in AAMT) is a permanent or temporary phenomenon. Nevertheless these three patients give grounds for cautious optimism and we would strongly recommend further trials to investigate this procedure. Both methods have produced improvement in the patients but only continued observation and assessment will determine which is the most satisfactory.

Acknowledgements

We are grateful to E. Bainbridge FRCS, Consultant Surgeon and W. Mitchell, X-ray, ward and theatre staff and secretaries for surgical, technical and essential support; the staff of the Robert Nursing Home for provision of foetal tissue and Dr. Desselberger (Regional Virology Laboratory) for HIV testing.

Veronica Turner typed the manuscript and checked the references.

References

1. Backlund EO, Granberg P, Hamberger B *et al* (1985) Transplantation of adrenal medullary tissue to striatum in parkinsonism. J Neurosurg 162: 169–172
2. Bakay RAE, Fuandaca MS, Barrow DL *et al* (1985) Preliminary report on the use of fetal tissue transplantation to correct MPTP-induced Parkinson-like syndrome in primates. Appl Neurophysiol 48: 358–361
3. Hitchcock E (1988) Recent experience with dopamine transplantation for Parkinson's disease. Proceeding of combined meeting Society of British Neurological Surgeons and Neurosurgical Society of Australia, Oxford, 14th April 1988. In press
4. Koller WC, Perlik S, Nausieda PA *et al* (1980) Drug holiday and management of Parkinson's disease. Neurology 30: 1257–1261
5. Lindvall O, Backlund EO, Farde L *et al* (1987) Transplantation in Parkinson's disease; two cases of adrenal medullary grafts to the putamen. Ann Neurol 22: 457–468
6. Madrazo I, Drucker-Colin R, Diaz V *et al* (1987) Open microsurgical autograft of adrenal medulla to the right caudate nucleus in 2 patients with intractable Parkinson's disease. New Engl J Med No 14, 316: 831–834
7. Madrazo I, Leon V, Torres C *et al* (1987) Transplantation of fetal substantia nigra and adrenal medulla to the caudate nucleus in two patients with Parkinson's disease. New Engl J Med No 1, 318: 51
8. Quinn NP (1987) Drug holiday. In: Koller WC (ed) Handbook of Parkinson's disease. Dekker, New York, pp 328–329
9. Redmond ED, Roth RH, Elsworth JD *et al* (1986) Fetal neuronal grafts in monkeys given methylphenyltetrahydropyridine. Lancet May 17, 1125–1127

Correspondence: Professor E. Hitchcock, Midland Centre for Neurosurgery and Neurology, Holly Lane, Smethwick, Birmingham, B67 7JX, U.K.

Pain

Acta Neurochirurgica, Suppl. 46, 53–57 (1989)
© by Springer-Verlag 1989

Clinical Use of Nociceptive Flexion Reflex Recording in the Evaluation of Functional Neurosurgical Procedures

L. García-Larrea[1], **M. Sindou**[2], and **F. Mauguière**[1]

[1] EEG Department Clin. Neurophysiology Unit, Hôpital Neurologique, Lyon, France, [2] Neurosurgery A Department, Hôpital Neurologique, Lyon, France

Summary

Nociceptive flexion reflexes (RIII) obtained by stimulation of sural nerve were studied in patients with intractable chronic pain before and after functional neurosurgery, either dorsal column stimulation (DCS, n = 15) or posterior selective rhizotomy in the dorsal root entry zone (DREZ, n = 5).

Dynamic study of RIII at supraliminal levels provided direct, quantitative and replicable evidence of the inhibition of nociceptive spinal reflexes by DSC. The effects of DSC on the RIII were highly correlated with subjective pain relief. In non-collaborative patients it was still possible to select the best DCS parameters (frequency, intensity) as those providing maximal RIII depression.

After posterior selective rhizotomy in the DREZ involving S 1-S 2 root levels postoperative evidence of selective extralemniscal lesioning could be assessed by the abolition or strong attenuation of nociceptive RIII, whereas preservation of the lemniscal pathways was evidenced by somatosensory evoked potentials.

Routine recording of nociceptive reflexes in man proved to be a useful tool for the objective evaluation of anatomo-physiological effects of functional neurosurgical procedures.

Keywords: Analgesic neurostimulation; microsurgical DREZ-tomy (MDT); dorsal column stimulation; nociceptive flexion reflex; transcutaneous neural stimulation; TENS; RIII.

Introduction

Spinal flexion reflexes have been studied both in normal human subjects[9, 14, 20, 22, 23] and in patients with various pathological conditions[5, 11, 14, 21, 24]. The observation by Willer that the threshold of one of these reflexes (RIII) was closely related to the subjective sensation of pain in man[20] prompted quantitative evaluation of these responses as a tool for human pain research. It has been demonstrated that, in normal subjects, both flexion RIII responses and pain sensation thresholds are similarly affected by pharmacological or behavioural manipulations[21, 22], and similar correla-tions between segmental reflexes and ascending pain signals have also been made in animals[3, 6]. Yet, most studies on flexion reflexes in pain syndromes have been limited to disclosing group-differences between patients and controls[21, 24]; no attempt has been made, to our knowledge, to investigate the utility of this non-invasive technique as an ancillary procedure for neurosurgical management of individual pain patients.

The aim of this work is to assess the clinical utility of nociceptive flexor reflex recording in patients undergoing two neurosurgical analgesic procedures: 1) Analgesic neurostimulation (epidural or transcutaneous) and 2) Microsurgical Rhizotomy at the Dorsal Root Entry Zone (Microsurgical DREZ-tomy). Analgesic neurostimulation was chosen because of its well established effects on pain signaling in animals[6, 7, 12, 13] as well as its still controversial long-term effectiveness in man[10, 19]. Microsurgical DREZ-tomy (MDT)[15, 16–18] provided an example of a functional procedure with the potential of selectively blocking nociceptive volleys without altering non-nociceptive signals. Our preliminary results suggest that the study of flexor responses in man cannot only provide relevant elements to our understanding of mechanisms involved in pain relief, but also effectively aid in the management of patients referred for neurosurgical procedures.

Patients and Methods

Twenty-five patients have been investigated; 18 were under analgesic neurostimulation, either epidural over the dorsal columns (DCS, n = 14) or transcutaneous (TENS, n = 4). The other 7 had undergone microsurgical rhizotomy at the dorsal root entry zone (microsurgical DREZ-tomy, MDT). All were fully informed of the aims and methods of the recording procedure and gave their consent.

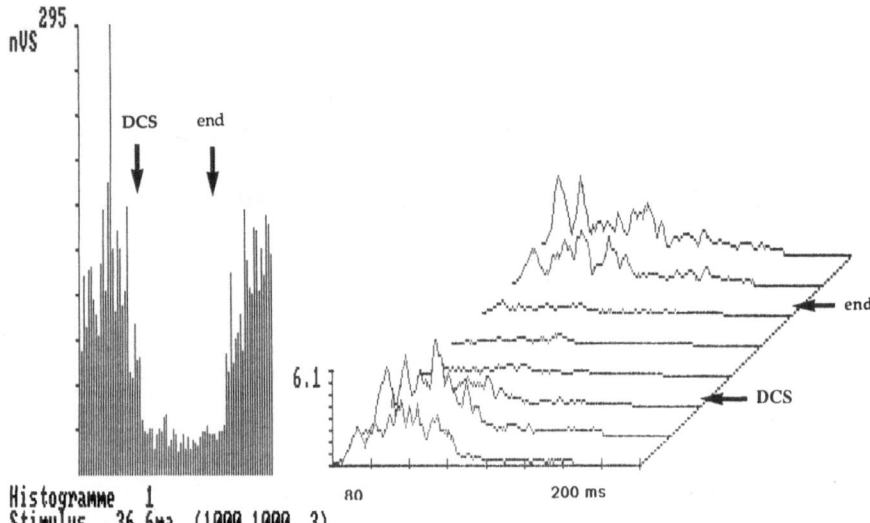

STIM MED PAR HEMIQUEUE GAUCHE POST-TRAUM   09-30-1987

Histogramme 1
Stimulus 36.6ma (1000 1000 3)

Fig. 1. DCS-related depression of flexion reflexes in a 35-year-old woman. At the right side of the figure, consecutive averaged reflexes during the analgesic neurostimulation session. Each trace is the rectified average of 5 single responses recorded at 15 sec intervals. The corresponding surface histograms of all reflexes elicited are shown on the left. Arrows indicate the beginning (DCS) and the end of neurostimulation, which was performed at T 10 level. Note that RIII, strongly depressed during dorsal column stimulation, regained basal values almost immediately after the end of the procedure

Patients lay comfortably on a bed or a reclinable arm-chair, in a semi-darkened room. The sural nerve was stimulated at the ankle and flexor reflexes were recorded over the short head of the ipsilateral biceps femoris, following the procedure described by Willer[20]. Each stimulus lasted 5 msec and consisted of a train of 3 constant current square pulses of 1 msec each, which were delivered at a rate of 500 Hz. The stimulator device could deliver intensities between 0 and 60 mA. Responses were amplified with a bandpass of 30–3,000 Hz (3 dB down, 6 dB/octave) and A/D converted at a rate of 200 Hz. They were stored on floppy disk for subsequent analysis. Individual and rectified-averaged traces, as well as surface histograms of the individual reflexes, were displayed on-line on the computer screen, enabling the examiner to follow the results of the test quantitatively during the recording session. Also, blocks of 5–8 responses corresponding to the same protocol status (for instance, with or without DCS) were averaged and the surface under the rectified trace computed for off-line statistical analysis.

To be identified as a flexor nociceptive (RIII) response the recorded reflex had to be polyphasic in shape with an onset latency between 80 and 130 msec after the stimulus. In addition, its presence had to be associated with a subjective sensation of "pricking" pain on the stimulation site, although this latter condition did not apply to patients with profound hypoesthesia of lower limbs. Special care was taken to avoid confusion with other flexion, non-nociceptive reflexes such as RII, which are usually biphasic in shape, with shorter latency and without accompanying subjective pain sensation. In all patients the response threshold was calculated as the minimum intensity evoking RIII reflexes with a 80–90% probability. Then a series of consecutive reflexes were obtained using a slightly suprathreshold stimulation intensity (usually at 1.5 times the threshold) to ascertain reproducibility.

In patients under analgesic neurostimulation (DCS or TENS) a longer series of 20–50 reflexes was obtained following a three-step protocol: (a) in basal conditions (before analgesic neurostimulation); (b) during neurostimulation and (c) immediately, and 10 minutes after the end of DCS/TENS.

In 12 patients the procedure was performed for each limb separately; in the others only the affected side was examined. Ten patients also underwent somatosensory evoked potentials recording to tibial nerve stimulation, either on the same day or as near as possible from the RIII recording session.

Results

1) Analgesic Neurostimulation (DCS or TENS)

A significant depression of lower limb nociceptive RIII reflexes during stimulation was observed in 10 patients (55.5%, see Table 1 A and Fig. 1). Eight had DCS (6 dorsolumbar, 2 cervical) and the other 2 brachial and popliteal TENS respectively. Depressed flexor reflexes tended to regain pre-stimulation levels shortly after cessation of DCS or TENS (Fig. 1). In one case only the RIII response remained significantly depressed more than 20 minutes after the end of a DCS session.

Seven out of 10 patients were initially satisfied with the pain-relieving effect obtained by the stimulating procedures. In two other patients pain was insufficiently relieved by neurostimulation at the beginning of the recording session, but adjustment of intensity and frequency of DCS so as to obtain maximal depression of RIII permitted a more effective control of spontaneous pains, which became already evident at the end of the recording session. In only one case was DCS-related depression of flexor reflexes definitely not associated with subjective pain relief. Since nociceptive reflexes showed a consistent and reproducible depression associated with DCS onset, malfunction of the

Table 1

A. DCS/TENS

		Pain relief	
		Satisfactory	Unsatisfactory
DCS (n=14)	RIII depressed (n=8)	7	1
	RIII unchanged (n=5)	2*	3
	RIII increased (n=1)	1	0

* Spastic paraparesis

		Satisfactory	Unsatisfactory
TENS (n=4)	RIII depressed (n=2)	2	0
	RIII unchanged (n=2)	0	2*

* One of them after 2 weeks of pain relief

B. MDT

Patient	MDT level	postoperative RIII	S.E.P.
M.J.	L4-S1	+	n.d.
R.C.	L2-S1	+	n.d.
M.L.	L4-S1	+	n.d.
D.B.	L1-S3	0	+
M.G.	S1-S3	0	+
J.C.	S1-S4	0	+
M.S.	S1-S4	+	0

stimulator device could be ruled out. This patient had been referred by her doctor with the suspicion of a psychogenic element to her pain, since DCS had abruptly ceased to be effective after a 2 years' period of good response.

RIII depression was found to be bilateral in 4 patients whose responses to each sural nerves were investigated separately. This includes 2 cases whose therapeutic stimulation was unilateral (one at the left popliteal fossa, the other over the right lumbar dorsal column).

2) Microsurgical DREZ-tomy (MDT)

Flexion reflexes after MDT at lumbosacral level were obtained in 7 patients. In 3 cases recordings were obtained before and after operation.

We defined RIII "abolition" as inability to elicit any nociceptive flexor response at a stimulus intensity of 50 mA, and under these conditions RIII flexion reflexes were abolished postoperatively in 3 cases. As shown in Table 1 B, at least the two first sacral roots must be involved in the microsurgical procedure for the postoperative abolition of nociceptive reflexes to sural nerve stimulation. Unfortunately, no preoperative RIII recording was available in these 3 cases, but all presented clinically exaggerated flexion reflexes preoperatively. In all of them it was still possible to demonstrate preserved dorsal column function postoperatively by somatosensory evoked potentials recording (Fig. 2). Similarly, one of these patients exhibited preserved RII reflexes (of cutaneous, non-nociceptive origin) whereas RIII responses were absent. In one case

SELECTIVE POSTERIOR RHIZOTOMY
(S1 – S4)

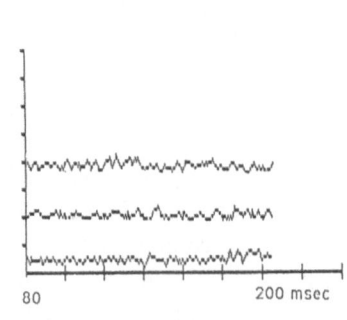

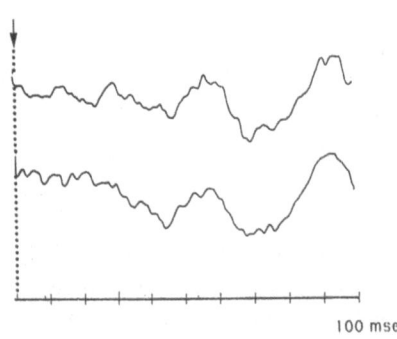

Fig. 2. Abolition of nociceptive flexion reflexes after bilateral S 1-S 4 DREZ-tomy for cancer pain. Flexion RIII responses were unobtainable even at maximal stimulation intensities (60 mA). It was nevertheless possible to record somatosensory evoked potentials to low intensity, lower limb stimulation, thus assessing the integrity of dorsal column system after operation

80   200 msec

RIII reflex (Algometry)

Train of stimuli
(3 x 1 msec, 60 mA)

100 msec.

Somatosensory evoked potentials

(0.2 msec, 10 mA)

RIII responses remained postoperatively despite a bilateral S 1-S 4 MDT. This 50-year-old woman had been operated upon for relief of spasticity secondary to multiple sclerosis and was ameliorated by surgery, although some degree of spasticity still persisted at the moment of RIII recording.

Discussion

Analgesic neurostimulation effectively depressed nociceptive flexion reflexes in 55% of patients, and in 77% of those who obtained satisfactory pain relief by the procedure.

The effects of DCS or TENS on flexion reflexes were largely independent of the stimulation site, being evident for lumbar, dorsal or cervical stimulation as well. This may be relevant to routine recordings since most patients can be effectively studied using a single segmental reflex (stimulation of sural nerve) whatever the location of their stimulating device. Inhibition of lower limb flexor nociceptive reflexes by distant DCS may be segmentally mediated by antidromic stimulation of dorsal columns[6, 7, 12] and/or involve supraspinal descending inhibitory mechanisms[4, 8]. Whatever the case, our limited experience indicates that RIII may no longer reflect the spinal inhibitory capacities of neurostimulation if significant cord lesions exist between the analgesic neurostimulation site and the segmental level of the evoked nociceptive reflex. This could explain the dissociation found in 2 patients (Table 1 A) whose reflexes to sural nerve stimulation remained unmodified under cervico-dorsal DCS in spite of a good clinical effect of neurostimulation. Since both patients were paraparetic, with lower limb hypesthesia, a cord lesion between DCS site and level of sural reflexes is likey, which could have prevented descending impulses to reach segments caudal to the lesion. In those cases other levels for eliciting RIII (superficial radial nerve, for instance) should be standardized.

The observed good correlation between RIII behaviour and effectiveness of neurostimulation is in keeping with studies in normals, which have shown that both nociceptive RIII and pain sensation can be similarly modulated by pharmacological or cognitive factors[22, 23]. This is also consistent with studies in animals showing that DCS simultaneously inhibits dorsal horn neurons and ascending volleys in anterolateral fasciculus[6]. All these data, along with our own, strongly suggest that neural circuitry subserving nociceptive flexion reflexes and ascending pain signals are closely related and probably coupled, and thus flexion reflex

recording could be an objective and non-invasive indicator of the inhibitory effects of analgesic neurostimulation in man.

This is also substantiated by two of our patients who were unrelieved, one week postoperatively, by DCS, but in whom a more effective analgesia developed when the intensity and frequency of the stimulator were adjusted in order to maximally depress the RIII reflex. Thus, nociceptive reflex recording could prove especially useful for patients in whom reliable clinical assessment of neurostimulation benefits is difficult, as in some patients with central pains secondary to stroke.

It is much too early to know whether, in the absence of cord lesions, the lack of effect of neurostimulation on RIII might predict its long-term inefficacy. If it were so, this could aid in eliminating superimposed placebo effects, as was probably the case in one of our patients who exhibited no variation in nociceptive reflexes under stimulation despite iterative activation of his brachial TENS. This patient, treated for postherpetic brachial pain, was clinically satisfied at the moment of RIII recording, but some weeks later analgesic stimulation ceased to be effective and the patient finally underwent cervical MDT.

Conversely to what was observed in neurostimulation, RIII abolition or depression following MDT was strongly dependent on the level of the operative lesions as was expected since that procedure aims at interrupting selectively A∂ and C afferents to the spinal cord[16–18] and thus involves the afferent branch of flexor reflexes activated by these fibers. Nociceptive reflexes to sural nerve stimulation were only found to be abolished postoperatively when MDT affected S 1-S 2 roots, and this restricts application of the technique in that multiple stimulation sites should be standardized before it can be used to evaluate postsurgical results in all operated patients.

MDT can be considered successful when it effectively interrupts the thinner, pain-signalling afferents at the dorsal root entry zone but preserves the overall input to the lemniscal system[16]. The selectivity of this procedure can now be ascertained by the coupling of nociceptive flexor reflex recording and of somatosensory evoked potentials, essentially conveyed by the dorsal column system.

Persistence of spinal flexor reflexes after MDT of the appropriate segmental levels might explain incomplete results observed in some patients undergoing this kind of surgery, as it was the case in our patient M. S. (Table 1 b). It would be logical to propose the recording of flexor reflexes intraoperatively to assess the completeness of the surgical lesion; however, this application may again be hampered by the fact that, in our experience, both intraoperative analgesia and intravenous barbiturates strongly depress RIII reflexes in humans. Further studies are needed to explore the clinical interest of RIII recording in MDT; our first data indicate however that this application is probably less straightforward than in the case of analgesic neurostimulation procedures.

Acknowledgements

Part of this work was conducted during a "GBM-TEP" procedure of prototype evaluation, supported financially by the french Departments of Research & Technology, Industry, Health and Social Security, and coordinated by the National Agency of Research Valorization (ANVAR), to whom the authors are indebted.

References

1. Bathien N, Bourdarias H (1972) Lower limb cutaneous reflexes in hemiplegia. Brain 95: 447–456
2. Besson JM, Chaouch A (1987) Peripheral and spinal mechanisms of nociception. Physiol Reviews 67: 67–186
3. Carstens E, Campbell IG (1988) Parametric and pharmacological studies of midbrain supression of the hind limb flexion withdrawal reflex in the rat. Pain 33: 201–213
4. Dickenson AH (1983) The inhibitory effects of thalamic stimulation on the spinal transmission of nociceptive information in the rat. Pain 17: 213–224
5. Dimitrijevic MR, Nathan PW (1967) Studies on spasticity in man: some features of spasticity. Brain 90: 1–30
6. Duggan AW, Foong FW (1985) Bicuculline and spinal inhibition produced by dorsal column stimulation in the cat. Pain 22: 249–259
7. Foreman RD, Beall JE, Applebaum AE, Coulter JD, Willis WD (1976) Effects of dorsal column stimulation on primate spinothalamic tract neurons. J Neurophysiol 39: 534–546
8. Gerhart KD, Yezierski RP, Fang ZR, Willis WD (1983) Inhibition of primate spinothalamic tract neurons by stimulation in ventral posterior lateral thalamic nucleus. Possible mechanisms. J Neurophysiol 49: 406–423
9. Hugon M (1973) Exteroceptive reflexes to stimulation of the sural nerve in normal man. In: Desmedt JE (ed) New developments in electromyography and clinical neurophysiology, vol III. Karger, Basel, pp 713–729
10. Keravel Y, Sindou M, Athayde A (1987) Indications of analgesic neurostimulation (TENS and spinal cord stimulation) in chronic neurological pain. In: Scherpereel *et al* (eds) The pain clinic II. VNU Science Press, pp 167–185
11. Kugelberg E, Eklund K, Grimby L (1960) An electromyographic study of the nociceptive reflexes of the lower limb. Mechanism of the plantar responses. Brain 83: 394–410
12. Lindblom U, Tapper N, Wiesenfeld Z (1987) The effect of dorsal column stimulation on the nociceptive response of dorsal horn cells and its relevance for pain suppression. Pain 4: 133–144
13. Lisney SJW (1979) Evidence for primary afferent depolarization of single tooth pulp afferents in the cat. J Physiol (Lond) 288: 437–447
14. Meink HM, Küster S, Benecke R, Conrad B (1985) Influence of stimulus parameters on reflex responses. Electroencephal Clin Neurophysiol 61: 287–298
15. Nashold JR BS, Ostdahl RH (1979) Dorsal root entry zone lesions for pain relief. J Neurosurg 51: 59–69
16. Sindou M, Quoex C, Baleydier C (1974 a) Fiber organization at the posterior spinal cord rootlet junction in man. J Comp Neurol 153: 15–26
17. Sindou M, Fischer G, Goutelle A, Mansuy G (1974 b) La radicellectomie postérieure sélective dans la chirurgie de la douleur. Neurochirurgie 20: 391–408
18. Sindou M, Fischer G, Mansuy G (1976) Posterior spinal rhizotomy and selective posterior rhizidiotomy. Prog Neurol Surg 7: 201–250
19. Wester K (1987) Dorsal column stimulation in pain treatment. Acta Neurol Scand 75: 151–155
20. Willer JC (1977) Comparative study of perceived pain and nociceptive flexion reflex in man. Pain 3: 69–80
21. Willer JC, Dehen H, Boureau F, Cambier J (1978) Further observations on the endogeneous morphine-like system in relation to congenital insensitivity to pain. J Med (NY) 9: 269–272
22. Willer JC, Boureau F, Albe-Fessard D (1979) Supraspinal influences on nociceptive flexion reflexes and pain sensation in man. Brain Res 179: 61–68
23. Willer JC, Bussel B (1980) Evidence for a direct spinal mechanism in morphine-induced inhibition of nociceptive reflexes in humans. Brain Res 187: 212–215
24. Willer JC (1986) Étude du seuil de la douleur par enregistrement des réflexes de flexion au cours des syndromes thalamiques. Rev Neurol 142: 303–307

Correspondence: L. García-Larrea, EEG Department, Hôpital Neurologique, 59 Bvd Pinel, F-69003 Lyon, France.

Acta Neurochirurgica, Suppl. 46, 58–61 (1989)
© by Springer-Verlag 1989

Intra-operative Spinal Cord Evoked Potentials During Cervical and Lumbo-sacral Microsurgical DREZ-tomy (MDT) for Chronic Pain and Spasticity (Preliminary Data)

D. Jeanmonod[1, 2], **M. Sindou**[1], and **F. Mauguière**[2]

Départements de [1]Neurochirurgie et de [2]Neurophysiologie Clinique, Hôpital Neurologique, Lyon, France

Summary

We have undertaken the intra-operative study of spinal cord surface evoked potentials in patients operated upon for pain and/or spasticity using the microsurgical DREZ-tomy (MDT) procedure. The goals of this work were 1) to collect data on spinal cord evoked potential components and 2) to analyze the effects of MDT on spinal cord physiology. The MDT consists of a therapeutic lesion in the ventro-lateral aspect of the dorsal root entry zone, directed to the activatory circuitry, and aiming at retuning the dorsal horn physiology towards inhibition. Averaged evoked potentials to peripheral nerve electrical stimulations were obtained from various loci on the surface of the dorsal columns of the cervical and lumbo-sacral spinal cord in 19 patients, using a small uninsulated silver ball electrode.

An initial far-field positivity was found, corresponding to a compound action potential in the proximal part of the brachial (or lumbo-sacral) plexus. Pre-synaptic compound action potentials were identified, most often composed of multiple successive sharp peaks. A post-synaptic field potential generated in the dorsal horn was recognized.

The MDT caused an immediate and irreversible decrease of amplitude down to a disappearance of the dorsal horn potential. This decrement was proportional to the amount of operated cord segments. In contrast, there has been a relative post-MDT sparing of the pre-synaptic action potentials originating from the operated cord segments, and the scalp contralateral parietal N 20 has been only reversibly affected by the therapeutic lesion. We thus argue for a specific involvement of dorsal horn physiology by the MDT, with a relative sparing of the dorsal column system.

Keywords: Chronic pain; DREZ; spasticity; spinal cord; evoked potentials.

Introduction

Whilst treating chronic pain and spasticity[10], we have studied intra-operative spinal cord surface evoked potential, or evoked electrospinogram (EESG), recordings during microsurgical DREZ-tomies (MDT) performed at the level of the cervical and lumbo-sacral

cord segments. MDT was introduced in 1972 on the basis of anatomical studies[11] of the human dorsal root entry zone (DREZ) which showed a topographical segregation of afferent fibers according to their size. It consists of a microsurgical lesion in the postero-lateral sulcus, penetrating the DREZ in its ventro-lateral aspect. It is supposed to interrupt predominantly the laterally-placed fine group III and IV fibers and destroy the medial excitatory part of Lissauer's tract. It should however spare the medially-placed lemniscal fibers and the lateral inhibitory part of Lissauer's tract. The global effect of MDT attempts a "retuning" of dorsal horn physiology towards inhibition.

The first goal of this study was to collect data on the EESG components. A large number of detailed studies were already available on spinal cord field potentials in animals[1, 3, 12], we developed for our human recordings, an approach as close as possible to these experimental conditions, for the sake of comparison. EESG studies in man are also available from intrathecal[7, 8], epidural[2, 4, 9] or skin and oesophageal derivations[5, 6]. The second goal of this work was the application of EESG recordings to the analysis of the effects of MDT on the physiology of the spinal cord.

Methods

Recordings were performed in 19 patients suffering either from chronic pain (12) or spasticity (7) and treated by MDT. The active EESG electrode was a silver ball measuring $750 \mu m–1$ mm long on $500–700 \mu m$ wide, placed on the dorsal column of the selected cord segments and maintained in position by a small cotton pad. The reference was always non-cephalic, the knee for lumbo-sacral cord studies, the shoulder for cervical studies, both contralateral to the peripheral nerve stimulation. Reference electrodes were subcutane-

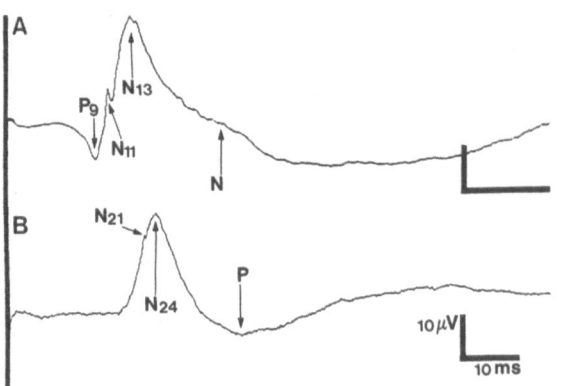

Fig. 1. EESG components. A) Recording from the surface of the dorsal column of segment C 7, ipsilateral to median nerve stimulation. The second later negative wave N 2 is labelled here N. B) Recording from the surface of the dorsal column of the L 5 cord segment, ipsilateral to tibial nerve stimulation

ous stainless steel needle electrodes 0.4 mm in diameter. Responses were recorded and averaged between 20 and 200 times by the RACIA EMG 21 P model (RACIA, Bordeaux, France), with a bin width of 137 µsec and analysis times of 60 (superior limb) and 90 (inferior limb) msec. The filter bandpass was set between 2 Hz and 2 kHz and the sensitivity was 100 µV/division. Bipolar stimulation was applied to the median nerve at the wrist, to the tibial nerve at the ankle and rarely in the popliteal fossa, using subcutaneous stainless steel needle electrodes. Monophasic square waves with a duration of 0.2 msec were delivered at a frequency between 4 and 6 Hz by a constant current isolated stimulator, at intensities just above motor threshold. Pre- and post-operative somatosensory evoked potentials were recorded on the skin surface over the C 6 or L 1 spinous processes or corresponding portion of the operative scar, and on the scalp.

The identified waves were labeled from their polarity and peak latency, and according to the data based on non-cephalic reference recordings in man[5,6]. In the text, the cervical potentials are mentioned first, and then followed by their lumbo-sacral counterparts between parentheses.

Results

A. Components of the EESG

Analysis of the EESG was performed on the basis of 122 recordings in the 19 patients of this series. Re-

cordings from sites, whose physiology might have been affected by the disease process, were interpreted with caution and in comparison with recordings in normal sites, which were available in approximately 40% of the cases. The present study has allowed identification of various EESG components, from the surface of the dorsal columns. These components are illustrated in Fig. 1, which shows typical examples of a cervical (A) and a lumbar (B) EESG. An initial positive wave P 9 (P 17) is identified. It is followed by a large slow negative wave, N 13 (N 24). The whole ascending slope of this negative wave has between 2 and 5 small and sharp peaks, which correspond to the surface N 11 (N 21) potential. Later phenomena include a second negative slow wave N 2 and a final very slow positive deflection, called P.

B. Effects of MDT on the EESG

The effects of MDT on the EESG have been documented in 44 recordings from 12 patients. Typical examples of these effects are shown in Fig. 2, in the lumbo-sacral (A and B) and cervical (C and D) spinal cord segments. After the MDT, the initial P 9 (P 17) wave is not only preserved, but often augmented. The N 13 (N 24) wave is suppressed when all stimulated cord segments are operated upon (Fig. 2 B), and decreased in proportion to these operated segments when some of them remain untouched (Fig. 2 D). To the contrary, the N 11 (N 21) potential is totally (Fig. 2 D) or relatively (Fig. 2 B) unaffected by the MDT. The later positive and negative waves are suppressed. Pre- and post-operative surface scalp and cervical recordings in a patient who had a C 5-C 8 MDT are displayed in Fig. 3. They show a disappearance of the cervical skin N 13 wave, examined at 7 days and 3 months postoperatively (Fig. 3 A–D). On the other hand, per-operative recordings after MDT (Fig. 3 F) show an acute latency shift

Fig. 2. Effects of MDT on the EESG. A) Recording from the surface of the dorsal column of segment L 5, ipsilateral to stimulation of the popliteal fossa. B) Same as A), but after MDT of all stimulated segments. C) Recording from the surface of the dorsal column of segment C 7, ipsilateral to median nerve stimulation. D) Same as C), but after C 5–C 8 MDT, thus sparing the T 1 cord segment

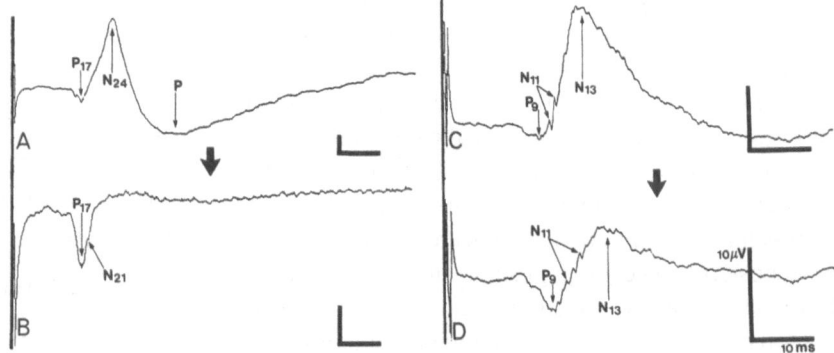

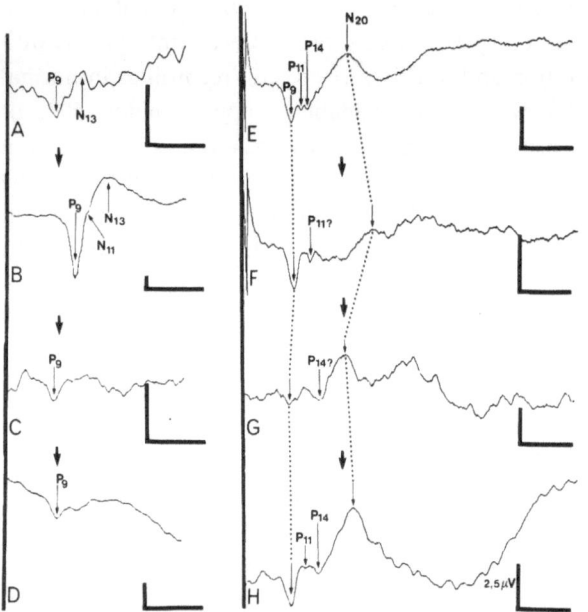

Fig. 3. A) Pre-operative skin recording on the spinous process of C 6 vertebra. Stimulation of the median nerve. B) Intraoperative recording from the surface of the dorsal column of segment C 6 with ipsilateral stimulation of the median nerve, after C 5–C 8 MDT, sparing the T 1 cord segment. C) Skin C 6 recording, same as A, but 7 days post-operatively. D) Same as C, but 3 months post-operatively. E) Pre-operative scalp parietal recording contralateral to median nerve stimulation. F) Same conditions as E, but intra-operatively, 10 minutes after the end of the MDT. G) Same as F, but 7 days post-operatively. H) Same as G, but 3 months post-operatively

and a loss of amplitude of the cortical N 20 wave, which is back to normal 7 days later (Fig. 3 G and H).

Discussion

The waveform, adjusted peak latencies and relative amplitudes of the EESG components described here are very similar to those found in animals in surface or depth recordings of spinal cord field potentials[1, 3, 12] and in other human studies[2, 4–9]. On this basis, the following conclusions can be made. The initial positive event P 9 (P 17) has been shown by skin and oesophageal recordings in man[5, 6] to be a far-field potential originating in the proximal part of the brachial (lumbo-sacral) plexus. The large slow negative wave N 13 (N 24) corresponds to the N 1 wave described in animals and related to post-synaptic spinal cord activity[1, 12]. The N 1 wave extends cranio-caudally further than the entry cord segments of the stimulated nerves, and has a negative posterior pole and a positive anterior one. Identical characteristics have been shown for the human N 13 and N 24 ([9], and authors unpublished data). The

emergence of the N 1 wave can be correlated to the activation of group I and II peripheral afferent fibers. The sharp peaks of the ascending slope of the N 13 (N 24) wave, also shown in other human studies[2, 4, 8], are the high resolution image of presynaptic successive axonal events, which become smoothed and reduced, or sometimes remain unconspicuous, in skin cervical or lumbar recordings[5, 6]. The second slow negative wave and the final, even slower, positive deflection correspond respectively to the experimental N 2 and P waves ([1, 12], and authors unpublished data). The P wave is the manifestation of pre-synaptic inhibition on primary afferent fibers. The N 2 wave has been shown to originate from a postsynpatic spinal cord activity consecutive to the activation of group II and large group III fibers.

Our recordings have shown that the MDT affects the N 13 (N 24) postsynaptic potential in proportion to the amount of stimulated cord segments which are operated upon, up to the time of its disappearance. Moreover, the later N 2 potential has been suppressed. If we accept the thesis that MDT does not destroy the dorsal horn itself but mainly interrupts its afferent fibers, then it can be argued that the operation not only interrupts fine group III and IV fibers but also involves at least some of the larger diameter axons, most probably where they turn ventrally to head for the dorsal horn. This effect is stable, the N 13 wave remaining absent at 3 months postoperatively. However, there was a relative preservation of the sharp presynaptic events (the N 11 and N 21 wave), and the acute intra-operative decrease and delay of the scalp N 20 wave were totally reversible, leading to a normal pattern 7 days later. This paralleled a major recovery of clinical discriminative and proprioceptive functions. We suggest therefore that the dorsal column-lemniscal system is spared, the partial reversible effects on it being attributable to operative manipulation and oedema, but not to surgical interruption. This suggests that the original goal of the MDT has been at least partially reached: a block of the activation of the pathological dorsal horn and a relative preservation of the discriminative and proprioceptive dorsal column system.

Acknowledgements

This work has received the generous financial support of the Foundation pour la Recherche Medicale, Paris, France. We want to acknowledge the very kind support of Dr. C. Fischer, the photographic help of Mr. S. Bello and the secretarial help of Mrs. D. Jeanmonod. The electrodes used here were developed and provided by Dr. C. Fischer, Département de Neurophysiologie Clinique, Hôpital Neurologique, Lyon, France.

References

1. Beall JE, Applebaum AE, Foreman RD, Willis WD (1977) Spinal cord potentials evoked by cutaneous afferents in the monkey. J Neurophysiol 40: 199–211
2. Beric A, Dimitrijevic MR, Prevec TS, Sherwood AM (1986) Epidurally recorded cervical somatosensory evoked potential in humans. Electroenceph Clin Neurophysiol 65: 94–101
3. Bernhard CG (1953) The spinal cord potentials in leads from the cord dorsum in relation to peripheral source of afferent stimulation. Acta Physiol Scand 29 [Suppl] 106: 1–29
4. Cioni B, Meglio M (1986) Epidural recordings of electrical events produced in the spinal cord by segmental, ascending and descending volleys. Appl Neurophysiol 49: 315–326
5. Desmedt JE, Chéron G (1981) Prevertebral (oesophageal) recording of subcortical somatosensory evoked potentials in man: the spinal P 13 component and the dual nature of the spinal generators. Electroenceph Clin Neurophysiol 52: 257–275
6. Desmedt JE, Chéron G (1983) Spinal and far-field components of human somatosensory evoked potentials to posterior tibial nerve stimulation analyzed with oesophageal derivations and non-cephalic reference recording. Electroenceph Clin Neurophysiol 56: 635–651
7. Makachinas T, Ovelmen-Levitt J, Nashold Jr BS (1988) Intraoperative somatosensory evoked potentials. A localizing technique in the DREZ operation. Appl Neurophysiol 51: 146–153
8. Nashold Jr BS, Ovelmen-Levitt J, Sharpe R, Higgins AC (1985) Intraoperative evoked potentials recorded in man directly from dorsal roots and spinal cord. J Neurosurg 62: 680–693
9. Shimoji K, Matsuki M, Shimizu H (1977) Wave-form characteristics and spatial distribution of evoked spinal electrogram in man. J Neurosurg 46: 304–313
10. Sindou M, Mifsud JJ, Boisson D, Goutelle A (1986) Selective posterior rhizotomy in the dorsal root entry zone for treatment of hyperspasticity and pain in the hemiplegic upper limb. Neurosurg 18: 587–595
11. Sindou M, Quoex C, Baleydier C (1974) Fiber organization at the posterior spinal cord-rootlet junction in man. J Comp Neurol 153: 15–26
12. Yates BJ, Thompson FJ, Parker Mickle J (1982) Origin and properties of spinal cord field potentials. Neurosurg 11: 439–450

Correspondence: Dr. D. Jeanmonod, Département de Neurochirurgie Pr M. Sindou, Hôpital Neurologique, 59 Bd Pinel, F-69003 Lyon, France.

Acta Neurochirurgica, Suppl. 46, 62–64 (1989)
© by Springer-Verlag 1989

Craniofacial Postherpetic Neuralgia Managed by Stereotactic Spinal Trigeminal Nucleotomy

J. R. Schvarcz*

School of Medicine, University of Buenos Aires, Buenos Aires, Argentina

Summary

Postherpetic craniofacial neuralgias are notoriously difficult to deal with. Nevertheless, stereotactic spinal trigeminal nucleotomy seems to be a rational approach, as both experimental and clinical data strongly suggest the relevance of nucleus caudalis for certain facial neurogenic pain phenomena.

From a series of 136 consecutive nucleotomies, 80 were performed for deafferentation pain. The long-term results of 25 such cases, who underwent this procedure for postherpetic neuralgia, are reported. Their pain was referred to the Vth, to the VII, IX and Xth, and to the C_{2-3} dermatomes. Abolition of the allodynia, and disappearance of, or marked reduction in, the deep background pain was achieved in 76% of the cases overall. The follow-up period ranged from 1 to 13 years. There were no untoward side-effects. Technical and electrophysiological data germane to accurate target placement are discussed.

Spinal trigeminal nucleotomy is then a specially suitable procedure for postherpetic craniofacial dysaesthesiae.

Keywords: Pain; stereotaxis; postherpetic neuralgia; trigeminal nerve; dorsal root entry zone; nucleotomy.

Introduction

Postherpetic facial neuralgia is still a neurosurgical challenge. It has certain peculiar characteristics. It does not occur in cases without a demonstrable sensory loss[9, 10]. The dysaesthetic pain, as well as other sensory aberrations, are typically more severe the greater the sensory deficit, which is often but not necessarily absolute[9, 10]. It can then presumably be equated to other deafferentation phenomena.

Surgical attempts to further interrupt the primary afferent neuron, at any level, including trigeminal tractotomy, are therefore unlikely to be successful[8, 22, 20] and, indeed, usually enhance rather than decrease the pain[5]. Contrarywise, lesions of the nucleus caudalis, *i.e.*, a lesion of the second order neurons, are a more a rational approach[12, 14, 16, 17].

Twenty-five such cases, who underwent this procedure, are reported.

Material and Methods

The technique has already been described[12, 16]. Briefly, patients are operated-on under local anaesthesia, with the head fully flexed within a modified Hitchcock's apparatus[2]. The spinal cord and caudal brain stem are outlined by water-soluble positive contrast. The spinal cord is then approached by a posterior route through the atlanto-occipital interspace.

Within the small confines of the mobile spinal cord, electrophysiological control is mandatory and can easily be achieved by impedance measurement, electrical stimulation and depth recording. Thus, the trigeminal region is neatly recognized between both the dorsal funiculus and the spinothalamic homunculi (Fig. 1).

From a series of 136 consecutive trigeminal nucleotomies, 80 were performed for deafferentation pain. Within this group, 25 patients had postherpetic neuralgia. Their pain histories ranged from 1 to 4.5 years. Their pain was referred to the first division in 23 cases, to the third in 1, to the VII/IX/Xth in 4, and to the C_{2-3}

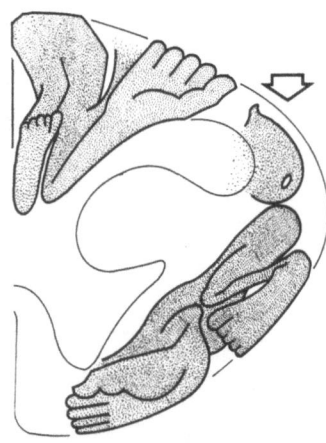

Fig. 1. Somatotopic organization of the high cervical spinal cord, in the light of stimulation data, demonstrating the trigeminal region (arrow)

dermatomes in 5 cases. They all had had extensive medical treatment without improvement and some had unsuccessful deep brain trial stimulation.

Results

Abolition of the allodynia, and a significant reduction in, or disappearance of, the deep background pain was achieved in 19 cases overall (76%). The follow-up period ranged between 1 and 13 years.

There were no untoward side-effects, although contralateral lumbosacral hypoalgesia, due to encroachment on the spinothalamic tract, sometimes occurred with large destructions of the ophthalmic area.

Discussion

It is possible to perform accurate stereotactic lesions of the fifth, of the seventh, ninth and tenth cranial nerves, as well as of the second and third cervical roots, at the spinal trigeminal nucleus.

Hitchcock[3] first performed a stereotactic trigeminal tractotomy in 1968, and he thereafter reviewed 21 cases with different sorts of intractable facial pain[4]. Hitchcock and I[5] reported, in 1972, radiofrequency lesions in the region of the descending trigeminal tract for postherpetic pain.

I have used this technique since 1971, naming the procedure trigeminal "nucleotomy"[12, 13], to emphasize the significance of lesioning primarily the second order neurons at the oral pole of the nucleus caudalis in certain neurogenic facial pain states[12-17]. Curiously enough, this target has recently been independently rediscovered[18].

Stereotactic nucleotomy shares some of the features of open surgical trigeminal tractotomy, but has some distinctive characteristics of its own. Thus, the results of this deeper, extensive nuclear lesion are in striking contrast to those of open medullary tractotomy, which so far has consistently failed to relieve postherpetic pain[8, 22, 20].

Black[1] has demonstrated that deafferentation by retrogasserian rhizotomy is gradually followed by a grossly abnormal, spontaneous neuronal hyperactivity at the nucleus caudalis, which is similar to that of an experimental epileptogenic focus. The time course of this spontaneous hyperactivity also paralleled the synaptic changes described by Westrum and Black[21] after retrogasserian rhizotomy. Similar changes have been reported in other chronically isolated neuronal populations of the spinal cord.

Since the nucleus caudalis represents the substantia gelatinosa at this level, this is a lesion of this structure.

The nucleus caudalis is a nodal point, where Kerr[6, 7] has demonstrated an extensive overlap between craniofacial and high cervical afferents. Also, an important ascending polysynaptic intranuclear pathway has been demonstrated[19].

This procedure then presumably removes the segmental pool of neuronal hyperexcitability and denervation hypersensitivity, eliminating convergence, and severing the ascending intranuclear pathways.

Thus, protracted abolition of the allodynia, and marked reduction in, or disappearance of, the deep background pain was achieved in 76% of the cases overall, followed up to 13 years.

Trigeminal nucleotomy is a safe and reasonably simple stereotactic technique, which allows accurate target placement by electrophysiological control prior to lesion making. It seems to be a specially suitable procedure for postherpetic craniofacial disaesthesiae.

References

1. Black R (1970) Trigeminal pain. In: Crue BL (ed) Pain and suffering. Thomas, Springfield, pp 119–137
2. Hitchcock ER (1969) An apparatus for stereotactic spinal surgery. Lancet i: 705–706
3. Hitchcock ER (1970) Stereotactic trigeminal tractotomy. Ann Clin Res 2: 131–135
4. Hitchcock ER (1978) Stereotactic spinal surgery. In: Carrea R (ed) Neurological surgery, with emphasis on non-invasive methods of diagnosis and treatment. Excerpta Medica, Amsterdam, pp 271–280
5. Hitchcock ER, Schvarcz JR (1972) Stereotactic trigeminal tractotomy for postherpetic facial pain. J Neurosurg 37: 412–417
6. Kerr FW (1970) The organization of primary afferents in the subnucleus caudalis of the trigeminal nerve. Brain Res 23: 147–165
7. Kerr FW (1975) Neuroanatomical substrates of nociception in the spinal cord. Pain 1: 325–356
8. Kunc Z (1970) Significant factors pertaining to the results of trigeminal tractotomy. In: Hassler R, Walker AE (eds) Trigeminal neuralgia. Thieme, Stuttgart, pp 99–100
9. Noordenbos W (1959) Pain: problems pertaining to the transmission of nerve impulses which give rise to pain. Elsevier, New York
10. Noordenbos W (1972) The sensory stimulus and the verbalization of the response: the pain problem. In: Somjen G (ed) Neurophysiology studied in man. Excerpta Medica, Amsterdam, pp 207–214
11. Schvarcz JR (1973) Stereotaxis of the spinal cord. In: International Congress Series 293. Excerpta Medica, Amsterdam, pp 75–76
12. Schvarcz JR (1974) Spinal cord stereotactic surgery. In: Sano K, Ishii S (eds) Recent progress in neurological surgery. Excerpta Medica, Amsterdam, pp 234–241
13. Schvarcz JR (1975) Stereotactic trigeminal nucleotomy. Confin Neurol (Basel) 37: 73–77

14. Schvarcz JR (1977) Postherpetic craniofacial dysaesthesiae. Their management by stereotactic trigeminal nucleotomy. Acta Neurochir (Wien) 38: 65–72
15. Schvarcz JR (1977) Functional exploration of the spinomedullary junction. In: Gillingham J, Hitchcock ER (eds) Advances in stereotactic and functional neurosurgery 2. Acta Neurochir (Wien) [Suppl] 24: 179–185
16. Schvarcz JR (1978) Spinal cord stereotactic techniques re trigeminal nucleotomy and extralemniscal myelotomy. Appl Neurophysiol 41: 99–112
17. Schvarcz JR (1979) Stereotactic spinal trigeminal nucleotomy for dysaesthetic facial pain. In: Bonica J, Liebeskind J, Fessard D (eds) Advances in pain research and therapy, vol 3. Raven Press, New York, pp 331–336
18. Schvarcz JR (1987) Letter to the editor. Neurosurgery 20: 348
19. Stewart WA, Stoops WL, Pillone PR, King RB (1964) An electrophysiologic study of ascending pathways from nucleus caudalis of the spinal trigeminal nuclear complex. J Neurosurg 21: 35–48
20. Sweet WH (1985) Personal communication
21. Westrum LE, Black RG (1968) Changes in the synapses of the spinal trigeminal nucleus after ipsilateral rhizotomy. Brain Res 11: 706–712
22. White JC, Sweet WH (1969) Pain and the neurosurgeon. A forty years experience. Thomas, Springfield

Correspondence: Prof. J. R. Schvarcz, M.D., Juncal 1845, Buenos Aires 1116, Argentina.

Acta Neurochirurgica, Suppl. 46, 65–66 (1989)
© by Springer-Verlag 1989

Spinal Cord Stimulation (SCS) in the Treatment of Postherpetic Pain*

M. Meglio, B. Cioni, A. Prezioso, and **G. Talamonti**

Istituto di Neurochirurgia, Università Cattolica, Roma, Italy

Summary

SCS is considered to be of poor value in treating postherpetic pain. We have retrospectively analyzed the results obtained in 10 patients suffering from postherpetic neuralgia. An epidural electrode was implanted, aiming the tip in a position where stimulation could produce paraesthesiae over the painful area. At the end of the test period 6 out of 10 patients reporting a mean analgesia of 52.5% underwent a permanent implant. At mean follow-up (15 months) all the 6 patients were still reporting a satisfactory pain relief (74% of mean analgesia). These figures remained unchanged at the next follow-ups (max 46 months). The result of SCS in our patients, although positive in only 60% of them, are remarkably stable with time. We therefore recommend a percutaneous test trial of SCS in every case of postherpetic neuralgia resistent to medical treatment.

Keywords: Spinal cord stimulation; postherpetic pain.

Introduction

Loeser[2] has recently pointed out that "we do not have any proven treatment program for postherpetic neuralgia and that ... the treatments utilized at the present time should have a very low risk of damaging the patient, for they all have only a small chance of providing long-term benefit". Spinal cord stimulation (SCS) even though an invasive procedure carries very low risk for the patient. Therefore, we utilize this procedure before considering any form of ablative surgery, in spite of the fact that discouraging results on this indication have been reported[1, 4].

In this report we analyze the results obtained in a group of patients with postherpetic pain.

Material and Methods

Fifteen patients (5 males and 10 females, age 49 to 79 years) were referred to us presenting with a history ranging from 4 to 16 months of postherpetic pain located in the trunk. They all had tried

* Supported by MPI, CNR progetto finalizzato: controllo dolore SP 8 and Vivian Smith Found.

several different non-invasive procedures and they were all on antidepressive medication. SCS was applied by means of percutaneously inserted epidural electrodes[3] connected to a percutaneous extension for test stimulation period and later to a permanent device (RF in 6 and ITREL in 4) for chronic stimulation.

The electrodes were positioned in the dorsal epidural space where comfortable paraesthesiae could be obtained in the painful area. Stimulation parameters were 85 c/sec, 0.2 msec, and with an intensity sufficient to produce paraesthesia. Stimulation was applied 20–30 min twice or three times daily for patients wearing the RF system and 64 sec every 1–4 min for those wearing an ITREL system.

The results are reported in terms of a percentage of analgesia (0% no pain relief and 100% complete pain relief) evaluated on the basis of patients' report on the visual analogue scale and with regard to their need for medication. Only the reduction of more than 50% of the original pain was regarded as a satisfactory result justifying a continued treatment.

Results

At the end of the test period 10 of the 15 patients reported a mean analgesia of 82.5% and were selected for chronic stimulation.

After three months of stimulation, analgesia showed

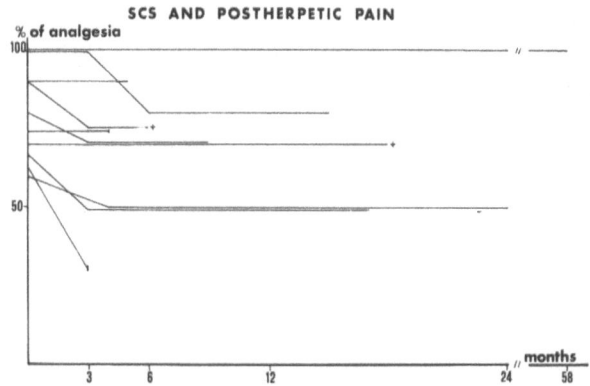

Fig. 1. Percentage of analgesia reported by each patient at successive follow-ups. Two patients died (+) and one stopped treatment due to mental illness

a slight decrease, but at the successive follow-ups pain relief remained unchanged (Fig. 1). SCS was discontinued after three months in one patient who showed signs of mental deterioration. In the remaining 9 patients SCS was the only treatment utilized at the latest follow-up (min 3, max 58 months; mean 15.9).

Complications occurred in 2 cases (failures of the system).

Discussion

SCS does not produce complete and permanent suppression of pain. Moreover, it does not cure postherpetic neuralgia. In fact, if one asks these patients about their pain they still complain, but it is quite obvious that their suffering is reduced. None of the patients would discontinue the use of the stimulator claiming that it provided more than 50% pain relief. They found the discomfort of the surgery required to implant the stimulation system well worthwhile and willingly came for revision as soon as there were changes of paraes-

thesiae or any other technical problems. It must be stressed that the analgesic effect was remarkably stable with time.

Taking into consideration the low risk of the procedure we believe that the results obtained in this series justify the policy of utilizing SCS as a first choice in the surgical treatment of postherpetic pain.

References

1. Erikson DL, Long DM (1983) Ten year follow-up of dorsal column stimulation. In: Bonica JJ *et al* (eds) Advances in pain research and therapy, vol 5. Raven Press, New York, pp 583–590
2. Loeser JD (1986) Herpes zoster and postherpetic neuralgia. Pain 25: 149–164
3. Meglio M, Cioni B, D'Amico E, Ronzoni G, Rossi GF (1980) Epidural spinal cord stimulation for the treatment of neurogenic bladder. Acta Neurochir (Wien) 54: 191–199
4. Nielson KD, Adams JE, Hosobuchi Y (1975) Experience with dorsal column stimulation for relief of chronic intractable pain. Surg Neurol 4: 148–152

Correspondence: M. Meglio, Istituto di Neurochirurgia, Università Cattolica S. Cuore, Largo A. Gemelli 8, I-00168 Roma, Italy.

Acta Neurochirurgica, Suppl. 46, 67–68 (1989)
© by Springer-Verlag 1989

CT-guided Percutaneous Cordotomy*

Y. Kanpolat, H. Deda, S. Akyar, and **S. Bilgiç**

University of Ankara, Ibni Sina Medical Center, Department of Neurosurgery and Radiology, Ankara, Turkey

Summary

Percutaneous cordotomy is a commonly applied and effective procedure among the ablative pain surgeries. As plain X-ray does not permit visualization of the target relative to the electrode the chances of obtaining good results are decreased and the risk of complications are increased.

The use of CT has been found to be useful in cordotomy. The procedure is performed under CT control on the patients who have previously been given 5 ml iohexol into the subarachnoid space. The needle electrode is manipulated by free hand technique. It is possible to measure the diameter of the spinal cord and to detect cord dislocation in the spinal canal. When the electrode system is introduced it is possible to visualize the tip of the electrode which is pushing or puncturing the spinal cord. As the procedure directly visualizes the relation of the electrode to the target it is possible to place the electrode in the lateral spinothalamic tract. Another advantage of the procedure is to enable us to visualize haematomas or other changes that may result from the cordotomy. The application of the technique and clinical results will be presented.

Introduction

One of the important problems in stereotactic ablative pain surgery is the visualization of the target electrode relation. Until recently, the relation of the target electrode has been visualized by the help of contrast medium[1, 7]. The necessity for direct visualization of target electrode relation in stereotactic pain surgery gave rise to the birth of the concept of "CT Guided Ablative Pain Surgery". CT guidance in pain surgery was first applied in extralemniscal myelotomy by us presented in Barcelona in 1987, and later published[3].

Material and Method

CT images from a 1200 SX device, 512 × 512 matrix, 3 mm slice thickness was used, and the quality of the image enhanced by di-

minishing the diameter of image formation. The patient is given 7 ml (240 mg) iohexol with LP into subarachnoid space 30 minutes prior to the procedure. The patient is laid on the CT table on supine position. Prior to the procedure, CT slices from the upper spinal cord are obtained to check whether the contrast medium spreads homogeneously. Meanwhile, the diameters of spinal cord measurements are determined. Following local anaesthesia, a classical cordotomy needle is introduced through C 1–C 2 space into the subarachnoid space. By drawing the lateral graph, the site of the needle is determined. Under the guidance of the axial slice images the direction of the needle is adjusted according to anteromedial or posterolateral site of anterolateral spinal cord (Fig. 1). After determining the target site the active electrode system is introduced, and new axial slices are obtained. Displacement of the spinal cord and/or electrode penetration is checked and the degree of penetration of the active tip of the electrode into the spinal cord is examined (Figs. 2–3) and the necessary corrections are made. Following measurement of impedence, and stimulation, R.F. lesion for the cordotomy is realized. New slices of the site of the operation are obtained postoperatively to visualize any possible complications.

This procedure has been used in six patients with intractable cancer pain; 4 of the cases had pulmonary carcinoma, one unilateral lumbar chondrosarcoma and another case had left ileo-pelvic bone metastases from hepatoma. There were no neurological complica-

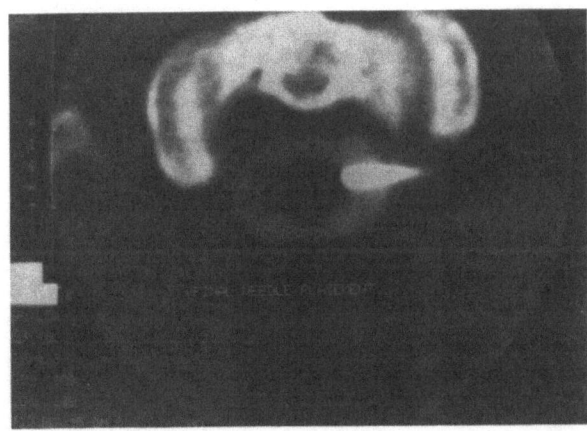

Fig. 1

* Dedication: We would like to dedicate this study to the memory of our dear master, tutor, man of science, the late Prof. Dr. Nurhan Avman.

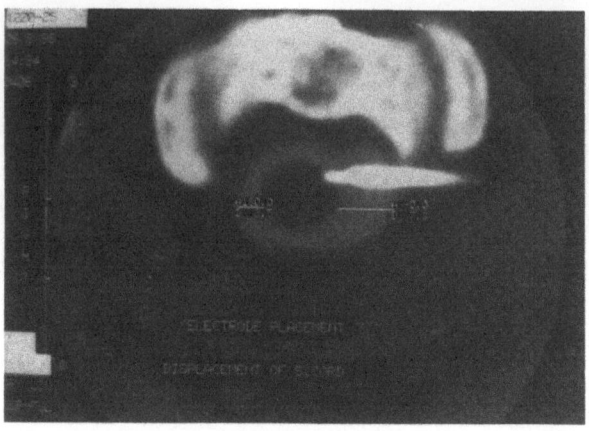

Fig. 2

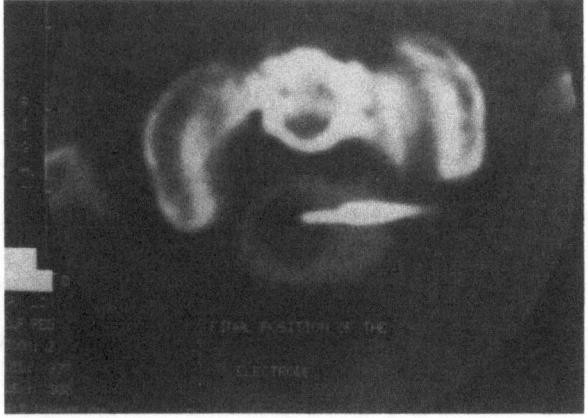

Fig. 3

tion. The Rosomoff* needle electrode system was used in four cases, in two cases Levin* cordotomy electrode system was used. Spinal cord displacement during the puncture is minimal with the Levin electrode. All cases had good analgesia and complete pain relief and three cases had segmental analgesia.

Although few cases have been subjected to the procedure, we believe it takes a shorter time than the classical cordotomy except in the first case.

Discussion

One of the most important problems in stereotactic ablative pain surgery is to demonstrate whether the electrode system has reached the target site or not. In classical application, the method applied gave this only through indirect visualization[5, 6]. We would like to emphasize that another most important problem in percutaneous ablative cord pain surgery is the mobility of the spinal cord[8], which can be displaced by the active electrode 5 mm without any puncture[8]. The visualization problem, and doubts about the penetration of the

---

* Radionics, Inc., Burlington, Massachusetts, U.S.A.

active electrode system have encouraged the use of impedance measurements, and neurophysiological recordings[2, 4]. Due to difficulties in the direct visualization of the target electrode relations, CT guided pain surgery was first used by us only for extralemniscal myelotomy. We are convinced of the usefulness of CT guidance as an imaging method for ablative pain surgery, especially in the spinal cord because of the direct visualization of the target electrode relation, and the diametric changes in the spinal cord before and during the procedure.

CT guided cordotomy has the following advantages:

The contrast material for the procedure is administered in the ward prior to the procedure. This provides both a more homogeneous distribution and shortens the period of surgery. It is possible to demonstrate the exact diameter of the spinal cord. With CT guidance it is possible to direct the needle to the selected sites of the anterolateral spinal cord. In this way, it is possible to localize the electrode in the anteromedial and posterolateral sites of lateral spinothalamic tract. It is possible to demonstrate exactly when the electrode system has entered the spinal cord, how much it has displaced the spinal cord, and how much it has punctured the spinal cord during the procedure.

In conclusion, we believe that CT guidance method will prove to be a classical imaging technique in the ablative pain surgery for extralemniscal myelotomy, percutaneous cordotomy and trigeminal tractotomy.

References

1. Batzdorf U, Bentson JR (1983) Use of metrizamide for percutaneous cordotomy. J Neurosurg 59: 545–547
2. Crue BL *et al* (1972) Percutaneous stereotaxic radiofrequency trigeminal tractotomy with neurophysiological recordings. Confin Neurol 34: 389–397
3. Kanpolat Y *et al* (1988) CT-guided extralemniscal myelotomy. Acta Neurochir (Wien) 91: 151–152
4. Kanpolat Y *et al* (1973) Experiments in percutaneous cordotomies by impedance method. AÜTFM, vol XXVI, II, pp 386–396
5. Onofrio BM (1971) Cervical spinal cord and dentate delineation in percutaneous radiofrequency cordotomy at the level of the first to second cervical vertebrae. Surg Gynecol Obstet 133: 30–34
6. Rosomoff HL *et al* (1965) Percutaneous radiofrequency cervical cordotomy. Technique J Neurosurg 23: 639–644
7. Smith R (1973) Outlining the cervical spinal cord with tantalum powder: application to percutaneous cordotomy. J Neurosurg 38: 257–260
8. Taren JA (1969) Target physiologic corroboration in stereotaxic cervical cordotomy. J Neurosurg 30: 569–584

Correspondence: Dr. Y. Kanpolat, University of Ankara, Ibni Sina Medical Center, Department of Neurosurgery, Sihhiye, Ankara, Turkey.

Acta Neurochirurgica, Suppl. 46, 69–72 (1989)
© by Springer-Verlag 1989

Myelotomies for Chronic Pain

D. van Roost[1] and J. Gybels[2]

[1] Neurochirurgische Universitätsklinik, Bonn, Federal Republic of Germany, [2] Kliniek voor Neurologie en Neurochirurgie, Universitair Ziekenhuis Gasthuisberg, Leuven, Belgium

Summary

The literature on myelotomy for the treatment of chronic pain was reviewed and a total of 635 published cases scrutinized. Two main modes of myelotomy can be distinguished 1) a longitudinal commissural section tuned to the segmental pain level and 2) a focused central lesion, irrespective of considerations of the metameric pain distribution, mainly carried out at a high cervical level.

Of the longitudinal commissural myelotomy, a posteriorly restricted and a complete type can moreover be discerned. The pain relief decays with time after myelotomy of any kind. Central myelotomy scores better than complete commissural section for malignant pain in a statistically significant manner but its superiority over posterior commissurotomy cannot be statistically proven.

Except of a girdle-shaped hypo-algesia, which is expected after the section of the decussating spinothalamic fibers in a complete commissurotomy, other—irregular—patterns of hypo-algesia have been observed, especially after central myelotomy.

This unusual lesion, provoking unusual hypo-algesia patterns, together with phenomena like a preserved sharp-blunt-discrimination within the hypo-algesic area, points at a different sensory channel that might be severed in a central myelotomy as compared with an anterolateral chordotomy or a complete commissurotomy. This hypothesis is matched with recent physiological evidences.

Keywords: Myelotomy; commissural central cord lesion (CCL); pain; spinal central gray matter.

Introduction

Many different kinds of surgical interruptions of nociceptive pathways have been carried out in the past and still find application in recent times despite the availability of non-ablative reversible techniques for chronic pain relief. For intractable body pain the anterolateral cordotomy remains the best known operation. Besides its more recent elegant percutaneous implementation, its clear anatomical foundation on a "pain-conducting" spinothalamic pathway with a somatotopic arrangement, contributed to its success. The bilateral procedure, however, often turned out to be unsatisfactory for midline visceral or bilateral shoulder pain, not least because of limiting respiratory complications at a high cervical level. An alternative intervention that by one cut interrupts the spinothalamic fibers from both sides of the body, where they cross the cord, could avoid injury to the motor pathway of the ventral horn or the pyramidal tract, Greenfield inspired Armour[1] to perform the first operation of this kind in 1926, which has been called a commissural myelotomy. The incision was in fact made along the posterior septum of the cord, aiming at the obliterated central canal, thus sectioning the gray commissure and, having reached the anterior median fissure, sectioning the white commissure too. The level of the myelotomy was selected for the presumed metameric pain input, taking into account the slightly upward shifted crossing of the spinothalamic fibers. The depth of the myelotomy however was not agreed even among the operation's pioneers: whereas Armour and Putnam[11] virtually cleft the cord in two halves, Leriche[7] stopped his incision at the central canal. Three reasons led to a posterior limitation of the commissurotomy: 1) the concern of avoiding damage to the anterior spinal artery; 2) the anatomical uncertainty, whether the spinothalamic decussation was located anterior or posterior to the central canal; 3) the empirical evidence that complete commissurotomy yielded no better results than posterior commissurotomy by authors of larger series[14].

Apart from the posteriorly limited and complete commissural myelotomies, a third type of myelotomy, which can be termed "central", must be considered (see Fig. 1). It consists of a stereotactic lesion of the spinal commissure at the first cervical segment or cervicomedullary junction with a posterior approach. This

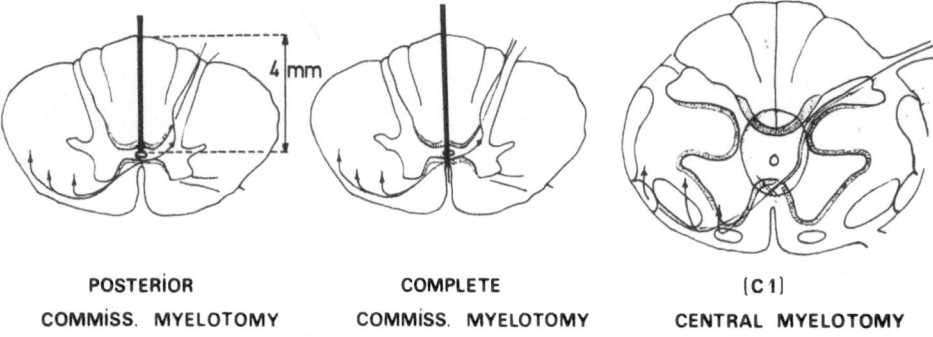

POSTERIOR   COMPLETE   (C 1)
COMMISS. MYELOTOMY   COMMISS. MYELOTOMY   CENTRAL MYELOTOMY

Fig. 1. Types of myelotomy for chronic pain

operation was first described in 1970 by Hitchcock[5]. The position of the electrode is mainly corroborated by the patient's reactions on electrostimulation, which characteristically produces chest or abdominal burning sensations or responses with a bizarre distribution. Compared with the complete and posterior commissural myelotomies, the central myelotomy displays 3 obvious differences: 1) a location, restricted to one, as a rule the most cephalad cord level, and not sited according to the metameric distribution of the pain; 2) a larger focal, particularly lateral extension; 3) a limited injury to the dorsal columns. Central myelotomy was found to produce irregular patterns of hypoalgesia and has been successfully applied to pain conditions—surprisingly—of the lower body half as well.

Methods and Material

Reviewing the literature, we found 28 authors or teams who reported their experiences with commissural myelotomy in a total of 635 published cases. Taking into account double entry of cases from updated and republished or from otherwise interrelated series (brackets in Table 1), an original number of 408 myelotomy cases remains, out of which 383 cases were followed up for at least 1 month. Those are listed in Table 1, classed according to the 3 myelotomy types and scrutinized for their successful results in benign versus malignant pain etiologies. Success means at least such an improvement that the patient no longer requests narcotic drugs and has no troublesome postoperative dysaesthesiae.

Results

Statistical analysis reveals a general decay of pain relief after myelotomy of any kind from 85% success immediately postoperatively to 70% after 1 month and to some 60% at the last follow-up ($\alpha = 0,01$). If only those cases with a follow-up exceeding 1 year are considered, 57% still show a good outcome. Central myelotomy surpasses complete commissural myelotomy re-

garding the overall outcome as well as regarding the relief of pain of malignant origin ($\alpha = 0,05$). Other statements cannot be statistically substantiated because of too small sample sizes. With the proviso of an insufficient number of posterior commissurotomies for reliable statistics, posterior commissurotomy does not seem to do worse than complete commissural or central myelotomy. A thorough analysis of the sensory effects showed that complete commissurotomy is followed—as expected—by a girdle-shaped hypoalgesia, although not in all cases. A girdle-shaped hypoalgesia does not occur at all after posterior commissurotomy, which can be understood from the anatomical site of the lesion, and only once after central myelotomy. Astonishingly all kinds of myelotomies may produce uncharacteristic hypoalgesia distributions, and the occurrence or type of hypoalgesia are not related to the pain relieving effect of a particular procedure, which is in striking contrast to the experience in anterolateral cordotomy. Several central-myelotomy cases have been described with good pain relief from areas where noxious stimuli could still be localized, though not perceived as painful, and where sharp/blunt-discrimination was preserved. The pain relieving and other sensory effects of central myelotomy, posterior commissurotomy and in part of complete commissurotomy too, therefore cannot be explained by segmental interruption of spinothalamic fibers on the mere analogy to anterolateral cordotomy.

Discussion

Over more than a century, experimental work has tried to elucidate the mechanism of nociception. The results of early behavioural studies after various types of cord section in non-human primates[9] turned out to be inconsistent with the later analgesic experiences of cordotomy in man. Invoking experimental data from

Table 1. *Late Results of Commissural Myelotomy*

Author (first)	Year	Operation type / level	Late success total	benign	malignant	upp./low.body		Follow-up
Leriche	1936	post. T4-5	100% n= 1					.5 W
Mansuy	1944	post. T5-7	(80% n=20)					'long-term'
Jentzer	47-48	post. T + S	0% n= 1		0% n= 1	0/1		15 M
Pieri	1951	post. T3-4	100% n= 1		100% n= 1	1/1		'months'
Wertheimer	1953	post. T4-6	65% n=80					
Armour	1927	compl. T11-L3	–					'days'
Putnam	1934	compl. C1-3-T4	50% n= 2		50% n= 2	1/2		2 M
Guillaume	1945	compl. T2-5	67% n= 3	67% n= 3				8 M
Arutjunov	1952	compl. C-T/T-L	87% n=15		87% n=15	0/2	13/13	?
Lembcke	1964	compl. C2-5-T1	86% n= 7	100% n= 6	0% n= 1	0/1		>10 Y
Sourek	69+77	compl. C4-7/T7-L1	84% n=38					≥ 5 Y f.5
Grunert	1970	compl. C3-T1/T-L	55% n=11	50% n= 2	56% n= 9		5/9	2 M – 3 Y
Broager	1974	compl. T9-11	58% n=33					2-12 M
Lippert	1974	compl. T2-4/T-L-S	75% n= 8	67% n= 3	80% n= 5	1/1	3/4	2- 7 M
Cook	1977	compl. C4-T1/5-L1	20% n=10	0% n= 7	67% n= 3		2/3	≥ 2 Y f.7
	1984	compl. T4-7-11	50% n= 2		50% n= 2	0/1	1/1	6 M
King	1977	compl. T9-11-S1	70% n=10	33% n= 3	86% n= 7		6/7	> 4 M; >2 Y
Fascendini	1979	compl. T6-8-L1	17% n= 6		17% n= 6		1/6	≤ 3 M
Adams	1982	compl. segmental	(67% n=24)	(67% n= 3)	(62% n=21)	(0/2)	(13/19)	?
Esposito	1982	compl. Conus	0% n= 3		0% n= 3		0/3	3 M
Fink	1984	compl. Conus	100% n= 1	100% n= 1				> 7 M
Goedhart	1984	compl. S2-5	37,5 n= 8		37,5 n= 8		3/8	2½ M- 6 Y
Sweet	1984	compl. Conus	33% n= 9		33% n= 9		3/9	2 M- 2½ Y
Ballantine	1969	centr. T12-S2	100% n= 1		100% n= 1		1/1	5 M
Hitchcock	1970	centr. C0-1	(43% n= 7)		43% n= 7	0/2	2/2	1-4 M
	1977	centr. C0-1	80% n=25					> 3 M
Nádvornik	1974	centr. C0-1	–					11 D
Papo	1976	centr. C2-4	43% n= 7	100% n= 1	33% n= 6	1/3	1/3	5 W – 1 Y
Schvarcz	1976	centr. C0-1	(82% n=45)	(80% n=10)	(83% n=35)			5 M- ≥9 M
	1978	centr. C0-1	76% n=75	64% n=14	78% n=61			6 M – 4 Y
	1984	centr. C0-1	(? n116)		(76% n=79)		(60/79)	½ – 30 M
Eiras	1980	centr. C0-1	58% n=12		58% n=12			2 – 22 M
Gildenberg	1984	centr. C1	0% n= 2		0% n= 2	0/2		> 2 M
		c.+ c. T9/10	66% n=12					> 2 M

animals for the explanation of phenomena in humans should therefore be done with caution. Earlier and more recent experiments with spaced hemisections on both sides of the cord[5, 2] in resp. cats and rats have shown some alteration but no abolition of nociception. These studies suggest repeated midline-crossings of some afferent pathway involved in pain signalling, which then is probably located near the cord commissure. In a fibre degeneration study after either antero-lateral cordotomy or myelotomy in monkeys, similar projections to the thalamic nuclei were found, without however clear somatotopic organization after myelotomy; a strikingly different topographical distribution of the projections to the periaqueductal gray and finally projections to the reticular substance of the medulla and pons, present in cordotomy but not in myelotomy material[6].

Only modern orthograde and retrograde tracing techniques allowed further investigation. These showed that neurones around the central canal in the rat contribute strongly to long ascending spinal cord projections, and that neurones within this area respond exclusively to noxious stimuli applied within small receptive fields[8, 10]. Moreover, the neurones around the central canal were found to closely resemble the cells of the marginal zone and the outer layer of the substantia gelatinosa by containing high levels of nociception-associated peptides like substance P and met-enkephalin[3, 12, 13].

The clinical data from myelotomy support a different pain processing modality within the spinal cord and the experimental data support the major role of the central cord gray matter in this respect, yet it remains unclear whether or not the central gray makes part of an independent sensory channel.

References

1. Armour D (1927) Lettsomian lecture on the surgery of the spinal cord and its membranes. Lancet April 2: 691–698

2. Basbaum AI (1973) Conduction of the effects of noxious stimulation by short-fiber multisynaptic systems of the spinal cord in the rat. Exp Neurol 40: 699–716

3. Gibson SJ, Polak JM, Bloom SR, Wall PD (1981) The distribution of nine peptides in rat spinal cord with special emphasis on the substantia gelatinosa and on the area around the central canal (lamina X). J Comp Neurol 201: 65–79

4. Hitchcock E (1970) Stereotactic cervical myelotomy. J Neurol Neurosurg Psychiatry 33: 224–230

5. Karplus JP, Kreidl A (1914) Ein Beitrag zur Kenntnis der Schmerzleitung im Rückenmark. Pflügers Arch Physiol 158: 275–287

6. Kerr FWL, Lippman HH (1974) The primate spinothalamic tract as demonstrated by anterolateral cordotomy and commissural myelotomy. In: Bonica JJ (ed) Advances in neurology, vol 4. Raven Press, New York, pp 147–156

7. Leriche R (1936) Du traitement de la douleur dans les cancers abdominaux et pelviens inopérables ou récidivés. Gazette des Hôpitaux (Paris) 109(51): 917–922

8. Martin GF, Vertes RP, Waltzer R (1985) Spinal projections of the gigantocellular reticular formation in the rat. Evidence for projections from different areas to laminae I and II and lamina IX. Exp Brain Res 58: 154–162

9. Mott FW (1892) Ascending degenerations resulting from lesions of the spinal cord in monkeys. Brain 15: 215–229

10. Nahin RL, Madsen AM, Giesler Jr GJ (1983) Anatomical and physiological studies of the gray matter surrounding the spinal cord central canal. J Comp Neurol 220: 321–335

11. Putnam TJ (1934) Myelotomy of the commissure, a new method of treatment for pain in the upper extremities. Arch Neurol Psychiat (Chicago) 32: 1189–1193

12. Schoenen J, Lotstra F, Vierendeels G, Reznik M, Vanderhaeghen JJ (1985) Substance P, enkephalins, somatostatin, cholecystokinin, oxytocin and vasopressin in human spinal cord. Neurology 35: 881–890

13. Schoenen J, van Hees J, Gybels J, de Castro Costa M, Vanderhaeghen JJ (1985) Histochemical changes of substance P, FRAP, serotonin and succinic dehydrogenase in the spinal cord of rats with adjuvant arthritis. Life Sci 36: 1247–1254

14. Wertheimer P, Lecuire J (1953) La myélotomie commissurale postérieure. A propos de 107 observations. Acta Chir Belg 6: 568–574

References of Table 1 can be found in D. van Roost „Die kommissurale Myelotomie in der Schmerzbehandlung". Thesis (to be published)

Correspondence: D. van Roost, M.D., Neurochirurgische Universitätsklinik, Sigmund-Freud-Strasse 25, D-5300 Bonn-Venusberg, Federal Republic of Germany.

Tumours

Acta Neurochirurgica, Suppl. 46, 75–78 (1989)
© by Springer-Verlag 1989

Differential Diagnosis Between Tumoural and Non-Tumoural Intracranial Lesions in Children: a Stereotactic Approach

C. Munari[1,2], J. Rosler Jr.[2,3], A. Musolino[2], O. O. Betti[1,2], C. Daumas-Duport[2], O. Missir[2], and J. P. Chodkiewicz[2]

[1] INSERM U 97, Paris, France, [2] Hospital Sainte Anne, Paris, France, [3] Istituto Medico Antartida, Buenos Aires, Argentine

Summary

The rational management of intracranial lesions should be based on the exact definition of the nature of the lesions and, when it is possible, on their spatial definition. Since External Radiotherapy (ERT) and cytostatic therapy are not free of undue effects, especially in children, such treatments should be used only when a "sure" diagnosis is obtained. The aim of this paper is to study the results allowed by the Talairach's stereotactic methodology in children. During the period January 1979–December 1986, 704 stereotactic procedures including serial biopsies, were performed at the S. Anne Hospital in Paris. One hundred forty-eight procedures (21%) concerned 134 children (78 M; 56 F) aged from 2 to 16 years. The interval between the occurrence of the first clinical symptoms and the stereotactic procedures varied between 1 and 180 months (m: 24 m). Fifty-two (40%) had previous therapeutic procedures without precise diagnosis. The lesions were hemispheric in 46 (34%) and deep seated in 88 (66%). The serial stereotactic biopsies proved the existence of a non-tumoural lesion in 20 children (14.9%): (cryptic vascular malformation: 5, cortical dysplasia: 3, haematoma: 3, ischaemia: 1, granuloma: 1, degenerative pathology: 2, cicatrix: 2, post-ERT alterations: 1, arachnoidal cyst: 2). Four were in the brain stem. In 3 patients (2%), a precise diagnosis was not obtained. The CT scan characteristics of the 20 non-tumoural lesions did not permit to establish a precise differential diagnosis. The therapeutic management was adapted to the diagnosis, avoiding potentially dangerous procedures in the 20 non-tumoural lesions.

Keywords: Brain tumours; child tumours; CT scan; external radiotherapy; stereotactic biopsy; stereotactic surgery.

Introduction

Since the pioneer work of Spiegel *et al.*[32], stereotactic methodology appeared as a new tool for obtaining specimen of deep brain tissue. Stereotactic biopsy, initially used in patients with extrapyramidal syndromes[17] later became, an increasingly popular means for precisely defining the nature of brain lesions[14, 27, 28, 34], especially after the beginning of the CT scan era[1, 18, 19, 26]. Although the literature in the diagnosis of brain lesions in adults is abundant[5, 12], there are no papers focusing the results of the stereotactic approach in children until 1983[7]. The correct diagnosis of a brain tumour is mandatory in choosing the different treatments and even more important in non-tumoural lesions, for avoiding potentially dangerous therapies.

We present our stereotactic experience in 134 consecutive paediatric cases.

Material and Methods

Patients: During the period 1979–1986, 134 children (78 m; 56 f) (age: 2–16 years) underwent 148 stereotactic procedures (21% of 704 performed during this period). The commonest clinical symptomatology was intracranial hypertension or epileptic seizures. The delay between the clinical onset and stereotactic investigation was from 1 month to 15 years (mean: 24 months). All of them had at least one CT scan (repeated in 93) previously to the admission at the S Anne Hospital. MRI was performed in 11. The lesions were hemispheric in 46, on the midline in 42, in the basal ganglia in 25, thalamo-peduncular in 11 and in the brain stem in 8. Two patients had diffuse lesions.

Surgical procedures: All the procedures were done, under general anaesthesia, using the Talairach's apparatus and methodology, as described previously[23, 34, 36, 37, 38, 40].

Safe and useful intracerebral trajectories were choosen on the basis of: peroperative stereoscopic angiography (170 procedures in 134 patients); stereoscopic ventriculography (in $^{127}/_{134}$) and the mathematical reconstruction of the CT scan data in intracranial stereotactic space[25].

The serial biopsies were done with the Sedan Vallicioni, partially modified, instrument (external diameter: 2.4–1.7 mm) (length of the aspiration window: from 4 to 9 mm). One hundred and ninety tracks (1.48/patient; 1.34/procedure) obtained 798 brain specimen (5.95/patient; 5.39/procedure; 4.01/track).

In highly functional areas (as in the brain stem), the aspiration of the brain tissue follows measurement of electrical impedance[4, 6] and electrical stimulations by repeated single shocks.

Every specimen was immediately divided in two parts for peroperative and paraffin embedding examination[8, 9]. The histopathological data were classified according to the WHO classification[41], partially modified by Daumas-Duport. A stereo-EEG investigation[2, 3, 21, 40], was performed in 13 children with severe, drug resistant, partial, lesion related epilepsy.

Results

1. Type of Lesions

The lesions have been classified as follows:
tumours : 111
cortical dysplasia: 3
non-tumoural pathology: 17
unclear diagnosis: 3
In the tumoural pathology, there were 10 neuro-epithelial dysembryoplasia (D.N.E.). Among the non-tumoural pathology: cryptic vascular malformation (5), haematomas (3), ischaemia (1), inflammatory pathology (1), degenerative pathology (2), arachnoidal cyst (2), cicatrix (1), "chronic epileptic modification" (1), non-tumoural post-irradiation gliosis (1).

Uncertain diagnosis was: gliosis versus glioma (1) and "periphery of probably non-tumoural lesion" (2).

2. Stereotactic Neuroradiological Investigations

The plain skull X-ray was normal in 66% of cases. The angiographic stereotactic and stereoscopic study was apparently normal in 29% of cases. The ventriculography was normal in 10.5% of cases. Both, angiographic and ventriculographic examinations were normal in 2.9% of cases. The angiographic study was normal in $^3/_{17}$ cases with non-tumoural lesions. Seven out of 13 children with a normal ventriculography had a non-tumoural pathology. Ventriculography was normal in $^4/_5$ patients with a non-tumoural and non-vascular lesion.

3. CT Scan Data

An hypodense area was noted in 53.8%, with contrast enhancement in 28%. A spontaneous hyperdensity was present in 43% and was enhanced in 25%. Among the 34 hypodense, not enhanced lesions, we found 4 (11.7%) non-tumoural lesions and 5 D.N.E.

Five of the 58 spontaneously hyperdense, non-enhanced, lesions, were non-tumoural: two cryptic vascular malformations, two haematomas and one dysplasia. Three cryptic vascular malformations, spontaneously hyperdense were enhanced by the contrast.

4. Clinical Effects of the Stereotactic Procedure

An immediate improvement was noted in $^{41}/_{134}$: 38 with, at least partially, cystic lesion, and 3 with haematoma.

A transitory (< 72 hours) increase of a pre-existant motor deficit was noted in 3 children (2.2%). 2 children (1.4%) with pre-operative motor deficit were permanently impaired.

Discussion

Before deciding the most appropriate treatment for an intracranial lesions, it is helpful to define: the type of the lesion, its precise anatomic localization and boundaries, the anatomo-functional relationships between the lesion and the neighbouring structures, the lesions growth potential.

The most important differential diagnosis is between tumoural and non-tumoural lesions. Several studies show that non-tumoural lesions are far from unfrequent (Table 1). This differential diagnosis is particularly important in children, because their brain is extremely sensitive to External Radiotherapy (ERT) and cytostatic therapy[10, 11, 31]. The importance of a precise diagnosis prior to ERT or other treatment is often unappreciated, as stressed by Reigel *et al.*[30].

Since ERT, or other treatment, should not be given before a precise diagnosis, the problem is to find which is the best method for obtaining it.

The data of the CT scan in our cases confirm that the CT image "can show identical or similar appearances for lesions of biologically different nature"[27] in tumoural and non-tumoural lesions. Moreover, the CT scan data can hardly be considered as providing reliable prognostic informations: Piepmeier[29] considers that low grade hemispheric astrocytic tumours have a significantly longer survival when they are not contrast enhanced (and when the age is lower), and, Stroink *et al.*[33] consider that paediatric brain stem gliomas have better prognosis when they are isodense and contrast enhanced. We agree with Hood *et al.*[16] that correct histological diagnosis is mandatory for therapeutic management and evaluating the results. Open surgery, for histological diagnosis alone, does not seem the best solution, since the morbidity and the mortality are relatively high[15], and the diagnostic results often unreliable[33].

Table 1. *Non-Tumoral Lesions*

Ostertag *et al.*	1980	9,7%	of 302 cases
Mundinger	1985	17%	of 904 cases
Moringlane and Ostertag	1987	14,7%	of 401 cases
Munari *et al.*	1987	17%	of 180 cases
This study (in children)	1988	15,9%	of 134 cases

Stereotactic biopsy is a safe and reliable diagnostic approach in large series of adult patients[5, 26, 27, 28], for both hemispheres and brain stem[24]. Since Broggi et al.[7], several recent studies concern stereotactic biopsies in children[13, 16, 20, 22], confirming that in children, the morbidity rate is very low (1.4% in this series), even if compared with the adult data (varying from 0% of the Rennes group[5], to 5.9% of Lundsford and Martinez[19] without peroperative angiography). The rate of non-diagnostic stereotactic biopsies is also low, 0 of Broggi et al.[7], 1.4% of this series, 1.5% of Moringlane et al.[20], 5.8% of Godano et al.[13].

Thus, we consider that stereotactic biopsy is a useful and safe tool for obtaining a precise neuropathological diagnosis of intracranial lesions in children. This diagnosis seems particularly valuable in the not infrequent cases with non-tumoural lesions avoiding the risks of a blind therapeutic management.

References

1. Apuzzo MLJ, Sabshin JK (1983) Computed tomographic guidance stereotaxis in the management of intracranial mass lesions. Neurosurgery 12: 277–285
2. Bancaud J, Talairach J, Bonis A, Schaub C, Szikla G, Morel P, Bordas-Ferrer H (1965) La stéréo-électro-encéphalographie dans l'épilepsie. Masson et Cie, Paris, pp 321
3. Bancaud J, Talairach J, Geier S, Scarabin JM (1973) EEG et SEEG dans les tumeurs cérébrales et l'épilepsie. Edifor ed., Paris, pp 351
4. Benabid AL, Persat JC, Chirossel JP, Rougemont J, (de) Barge N (1978) Delimitation des tumeurs cérébrales par stéréo-impédo-encéphalographie (S.I.E.G.). Neurochirurgie 24: 3–14
5. Benabid AL, Blond S, Chazal J, Cohadon F, Daumas-Duport C et al (1985) Les biopsies stéréotaxiques des néoformations intra-crâniennes. Réflexions à propos de 3052 cas. Neurochirurgie 31: 295–301
6. Broggi G, Franzini A, Passerini A (1979) Correlations between impedance values and deep brain tumours during stereotactic biopsies. In: Szikla G (ed) Stereotactic cerebral irradiations. Elsevier, North Holland, pp 51, 56
7. Broggi G, Franzini A, Migliavacca F, Allegranza A (1983) Stereotactic biopsy of deep brain tumours in infancy and childhood. Child's brain 10: 92–98
8. Daumas-Duport C, Monsaingeon V, Szenthe L, Szikla G (1982) Serial stereotactic biopsies: a double histological code of gliomas according to malignancy and 3 D configuration, as an aid to therapeutic decision and assessment of results. Applied Neurophysiology 45: 431–437
9. Daumas-Duport C, Monsaingeon V, Blond S, Munari C, Musolino A, Chodkiewicz JP, Missir O (1987) Serial stereotactic biospies and CT scan in gliomas: correlative study in 100 astrocytomas, oligo-astrocytomas and oligo-dendrocytomas. J Neuro-oncol 4: 317–328
10. Dickinson WP, Berry DH, Dickinson L et al (1978) Differential effects of cranial radiation on growth hormone response to arginine and insulin infusion. J Ped 92: 754–757
11. Duffner PK, Cohen ME, Thomas PRM, Lansky SB (1985) The long-term effects of cranial irradiation on the central nervous system. Cancer 56: 1841–1846
12. Edner G (1981) Stereotactic biopsy of intracranial space occupying lesions. Acta Neurochir (Wien) 57: 213–234
13. Godano V, Frank F, Fabrizi A, Frank Ricci R (1987) Stereotaxic surgery in the management of deep intracranial lesions in infants and adolescents. Child's Nerv Syst 3: 85–88
14. Heath RG, John S, Foss O (1961) Stereotactic biopsy: a method for the study of discrete brain regions of animals and man. Arch Neurol (Chicago) 4: 291–300
15. Hitchcock E, Sato F (1964) Treatment of malignant gliomata. J Neurosurg 21: 497–505
16. Hood TW, Venes JL, McKeever PE (1986) Stereotactic biopsy of paediatric brain stem tumours. J Paediatr Neurosci 2: 79–88
17. Housepian EM, Pool JL (1960) The accuracy of human stereo-encephalotomy as judged by histological confirmation of roentgenographic localization. J Nerv Ment Dis 130: 520–525
18. Leksell L, Jernberg B (1980) Stereotaxic and tomography. A technical note. Acta Neurochir (Wien) 52: 1–7
19. Lundsford D, Martinez J (1984) Stereotactic exploration of the brain in the Era of computed tomography. Surg Neurol 22: 222–230
20. Moringlane JR, Graf N, Ostertag CB (1987) Papel de la exploracion estereotaxica en el diagnostico y tratamiento de los tumores cerebrales del nino. Neurologia (Esp) 2, 5: 202–210
21. Munari C, Bancaud J (1985) The role of the stereo-EEG in the evaluation of partial epileptic seizures. In: Morselli PL, Porter RJ (eds) The epilepsies. Butterworth, London, pp 267–306
22. Munari C, Musolino A, Franzini A, Demierre B, Betti O, Daumas-Duport C, Broglin D, Missir O, Chodkiewicz JP (1986) Esplorazione stereotassica con biopsie seriate nelle patologia tumorale infantile. Boll Inf Soc It Neurochir 1: 252
23. Munari C, Musolino A, Demierre B, Betti O, Franzini A, Rosler JR, Broglin D, Daumas-Duport C, Missir O (1987) Complémentarité de la tomodensitométrie, des biopsies stéréotaxiques étagées et de la stéréo-électro-encéphalographie (SEEG) dans la définition spatiale des lésions intracérébrales. Rev Electroencephalogr Neurophysiol Clin 17: 3–10
24. Munari C, Musolino A, Rosler JR, Blond S, Demierre B, Betti OO, Daumas-Duport C, Missir O, Chodkiewicz JP (1987) Stereotactic approach to space occupying lesions in the posterior fossa. Appl Neurophysiol 50: 200–202
25. Musolino A, Munari C, Betti O, Landre E, Broglin D, Demierre B, Missir O, Daumas-Duport C, Chodkiewicz JP (1987) Intérêt et technique du transfert des données tomodensitométriques dans les coordonnées stéréotaxiques du système Talairach. Rev Electroencephalogr Neurophysiol Clin 17: 11–24
26. Mundinger F (1985) CT stereotactic biopsy for optimizing the therapy of intracranial processes. Acta Neurochir (Wien) [Suppl] 35: 70–74
27. Ostertag GCB, Mennel HD, Kiessling M (1980) Stereotactic biopsy of brain tumours. Surg Neurol 14 (4): 275–283
28. Pecker J, Scarabin JM, Brucker JM, Vallee B (1979) Démarche stéréotaxique en neurochirurgie tumourale. Pierre Fabre ed, Paris, 301 pp
29. Piepmeier JM (1987) Observations on the current treatment of low grade astrocytic tumours of the cerebral hemispheres. J Neurosurg 67: 177–181
30. Reigel DH, Scarff TB, Woodford JE (1979) Biopsy of pediatric brain stem tumors. Child's Brain 5: 329–340

31. Rowland J, Glidewell O, Sibley R, Holland JC, Brecher M, Tull B *et al* (1982) Effect of cranial radiation on neuropsychologic function in children with acute lymphocytic leukaemia. Asco Abstracts, March, p 123

32. Spiegel EA, Wycis HT, Marks M, Lee AJ (1947) Stereotaxic apparatus for operations on the human brain. Science 106: 349–350

33. Stroink AR, Hoffman HJ, Hendrick EB, Humphrey RP (1986) Diagnosis and management of paediatric brain stem gliomas. J Neurosurg 65: 745–750

34. Szikla G, Peragut JC (1975) Irradiation interstitielle des gliomes. In: Constans JP, Schlienger M (eds) Radiothérapie des tumeurs du système nerveux central. Neurochirurgie [Suppl] 21: 187–228

35. Szikla G, Bouvier G, Hori T, Petrov V (1977) Angiography of the human brain cortex. Springer, Berlin Heidelberg New York, pp 273

36. Szikla G, Blond S, Daumas-Duport C, Missir O, Miyahara S, Munari C, Musolino A, Schlienger M (1983) Stereotaxis in management of brain tumors: three dimensional angiography, sampling biopsies and focal irradiation using the Talairach stereotactic system. Ital J Neurol Sci [Suppl] 2: 83–96

37. Talairach J, Hecaen H, David M *et al* (1949) Recherches sur la coagulation thérapeutique des structures sous-corticales chez l'homme. Rev Neurol 81: 4–24

38. Talairach J, Aboulker J, Ruggiero G, David M (1954) Utilization de la méthode radiostéréotaxique pour le traitement radioactif *in situ* des tumeurs cérébrales. Rev Neurol 90: 656–657

39. Talairach J, David M, Tournoux P, Corredor H, Kvasina T (1957) Atlas d'anatomie stéréotaxique des noyaux gris centraux. Masson et Cie ed, Paris

40. Talairach J, Bancaud J, Szikla G, Bonis A, Geier S, Vedrenne C (1974) Approche nouvelle de la neurochirurgie de l'épilepsie. Méthodologie stéréotaxique et résultats thérapeutiques. Neurochirurgie [Suppl] 10: 1–240

41. Zülch KJ (1979) Types histologiques des tumeurs du S.N.C. OMS (ed), Geneve

Correspondence: Dr. C. Munari, INSERM U 97, 2ter rue d'Alésia, F-75014 Paris, France.

Acta Neurochirurgica, Suppl. 46, 79–81 (1989)
© by Springer-Verlag 1989

Natural History of Neuroepithelial Tumours: Contribution of Stereotactic Biopsy

R. Roselli, M. Iacoangeli, M. Scerrati, and **G. F. Rossi**

Institute of Neurosurgery, Catholic University, Rome, Italy

Summary

Quantitative tumour growth into the brain (stage of the disease) and qualitative tumour evolution in the time (progression) are the two basic aspects of the natural history of the cerebral neoplastic disease. Recently the development of neuroradiological imaging (CT and MR) and the progress in biopathology of the nervous system tumours introduced new concepts like heterogeneity of neuroepithelial tumours or evolution to anaplasia. The findings obtained in 159 neuroepithelial tumours studied with stereotactic biopsy from 1980 to 1987 are presented. Most of them were glioblastomas (n = 43; 27%) and astrocytic tumours (n = 81; 50.9%). Twenty-nine cases of fibrillary astrocytomas (35.8% of all astrocytic tumours) showed focal anaplasia (progression). In 10 out of the 43 glioblastomas (23.3%) signs of astrocytic differentiation were clearly evident (secondary glioblastoma?).

Our data confirm that neuroectodermal tumours, particularly of astrocytic series, undergo progression through anaplasia, which may be at the beginning a circumscribed phenomenon (focal anaplasia).

The staging of the disease (tumour growth) into cerebral nervous system in some cases can not be correctly expressed through the neuroradiological imaging. Sometime CT scan may underestimate the true extension of the lesion. On the contrary, MR may overrate the extension of the lesion. Such mistakes in evaluation of tumour staging may be corrected through seriate stereotactic biopsy.

Keywords: Brain tumours; staging; progression; stereotactic biopsy.

Introduction

Natural history is of paramount importance in the knowledge of the biopathology and prognosis of neoplastic disease[1, 8, 11, 16–20]. It includes: 1) quantitative tumour growth into the involved organ (stage of the disease)[1, 14, 15]; 2) qualitative tumour evolution in the time, namely the increase of malignancy (progression) related to the appearance or accentuation of heterogeneity and by the evolution towards anaplasia[10]. In neurosurgery the detection of the last events has been encouraged recently by the progress of neuroradiological imaging techniques (CT and MRI) and of bio-

pathology. It is important to correlate the non-homogeneous areas (different density or intensity) in the neuroradiological pictures of some neuroepithelial tumours (in particular those of the astrocytic line), to the phenotypic heterogeneity (focal anaplasia) found histologically[1–7]. Because of its spatial precision, the stereotactic biopsy appears the right means to reach this goal[2, 3], and our experience in cerebral tumours of the astrocytic line is reported.

Material and Methods

One hundred and eighty intracranial tumours underwent stereotactic biopsy between 1980 and 1987 in our Institute. Indication for biopsy was given on the basis of the neuroradiological investigations indicating an expanding lesion of dubious nature and for which the surgical approach may be dangerous[9]. One or more biopsy tracks were planned and performed according to the characteristics of the neuroradiological images. Biopsy was carried out with the Talairach stereotactic instrument utilizing either orthogonal or polar approaches[13]. Specimens were taken stepwise along the chosen biopsy tracks usually at a distance of 5–10 mm.

Results

Out of the 180 intracranial biopsied tumours, 97 (53.3%) turned out to belong to astrocytic line: 4 pilocytic, 63 fibrillary and 30 anaplastic astrocytomas. Twenty-nine cases of fibrillary astrocytomas (30% of all the astrocytic tumours) showed focal anaplasia. The correlation between pathological and neuroradiological data showed the following:

1) unhomogeneous image on the CT scan was present in 59 cases which was related to focal anaplasia within an otherwise differentiated tumour in 29 cases (Fig. 1);

2) tumoural extension in hypodense astrocytomas cannot be properly detected by CT scan; in 34 cases biopsy showed that it was larger than the CT image;

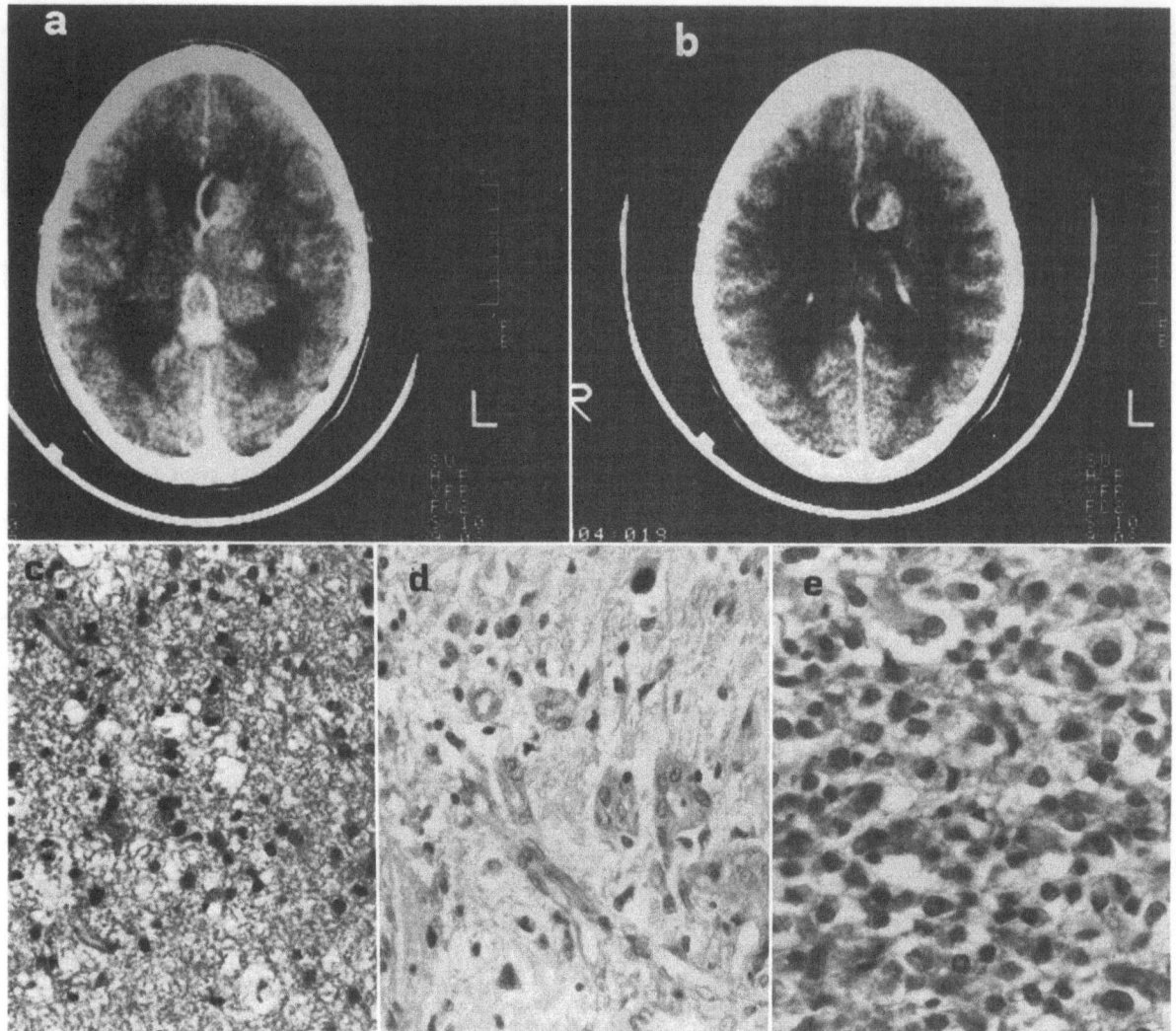

Fig. 1. CT scan of a deep seated astrocytic tumour of the left hemisphere (a) with an unhomogeneous enhanced area in its anterior part (b); bioptic samples showing a fibrillary astrocytoma (c) with an area of focal anaplasia (d, e) corresponding to the enhanced area on the CT scan

3) on the contrary in the 21 cases studied with MRI, the extension of the tumour, as verified histologically, was smaller then the MRI area of abnormal intensity (Fig. 2).

Discussion

Diagnosis and therapy are the main goals of the neurosurgeon in the management of the brain tumours. However the knowledge of the natural history of neuroepithelial tumours did not receive so far particular attention[10, 14]. The possibility given by stereotactic biopsy to precisely correlate the pathological findings with the neuroradiological characteristics, its safety and reliability[2, 3, 9, 12, 14] can offer significant information about staging and progression of primitive neuroepithelial tumours. Our data confirm that, in a relevant percentage (30%) of cerebral tumours of the astrocytic line, focal anaplasia is present. It seems quite likely that this finding indicate progression. This view is supported by unpublished data of signs of astrocytic differentiation in 10 out of 43 glioblastomas. The crucial proof can be obtained only through serial pathological controls, during the patients survival time. Some help might come, however, by a certain correlation between pathological data, as provided by stereotactic biopsy and neuroradiological findings[2, 3]. If that can be achieved instead of planning seriate biopsies the timing of the biological controls might be guided by serial neuroradiological studies.

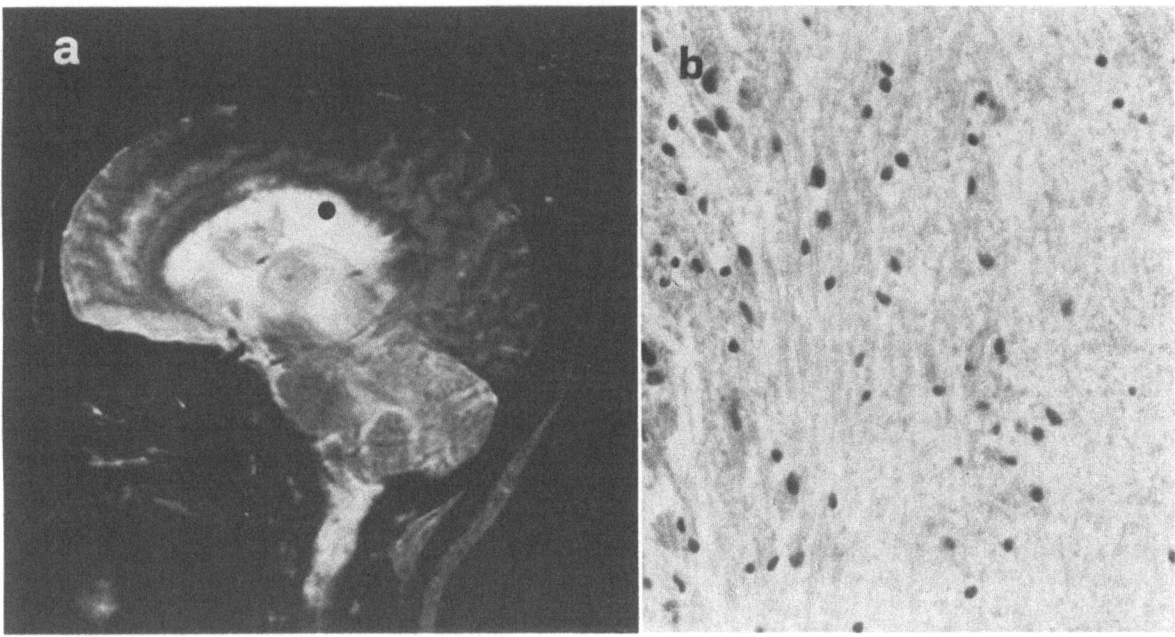

Fig. 2. a) MRI (T 2 weighted) of the same case of Fig. 1 and b) bioptic specimen taken from point indicated in a) showing negative results for tumoural tissue

Acknowledgements

This research is partially supported by a Grant of Ministry of Public Education of Italy.

References

1. American Joint Committee (1983) Cancer staging and end results reporting. Manual for staging of cancer. AJC Chapter 39: 219–226
2. Kelly PJ, Daumas-Duport C, Kispert DB, Kall BA, Scheithauer BW, Illing JJ (1987) Imaging based stereotaxic serial biopsies in untreated intracranial glial neoplasms. J Neurosurg 66 (6): 865–874
3. Kelly PJ, Daumas-Duport C, Scheithauer BW, Kall BA, Kispert DB (1987) Stereotactic histologic correlation of computed tomography and magnetic resonance imaging defined abnormalities in patients with glial neoplasms. Mayo Clin Proc 62 (6): 450–459
4. Kleihues P, Volk B, Anagnostopoulos J, Kiessling M (1984) Morphologic evaluation of stereotactic brain tumour biopsies. Acta Neurochir (Wien) [Suppl] 33: 171–181
5. Moschini M, Scerrati M, Colosimo C, Fileni A, Luna R, Roselli R, Rossi GF (1984) Integrazione tra neuroradiologia e neurochirurgia stereotassica nella diagnosi e nella definizione spaziale degli espansi endocranici. In: De Dominici R *et al* (eds) Radiologia-Firenze, Monduzzi-Bologna, pp 221–224
6. Ostertag CB, Mennel HD, Kiessling M (1980) Stereotactic biopsy of brain tumours. Surg Neurol 14: 275–283
7. Ostertag CB, Volk B, Shibata T, Burger P, Kleihues P (1987) The monoclonal antibody Ki-67 as a marker for proliferating cells in stereotactic biopsies of brain tumours. Acta Neurochir (Wien) 89: 117–121
8. Prodi G (1977) La biologia dei tumori. Ambrosiana, Milano
9. Rossi GF, Scerrati M, Roselli R (1987) The role of stereotactic biopsy in the surgical treatment of cerebral tumours. Appl Neurophysiol 50: 159–167
10. Rubistein LJ, Herman MM, Van den Berg SR (1984) Differentiation and anaplasia in central neuroepithelial tumours. In: Rosenblum ML *et al* (eds) Brain tumour biology. Progress in experimental tumour research, vol 27. Karger, Basel New York, pp 32–48
11. Russel DS, Rubistein LJ (1971) Pathology of tumours of the nervous system. Arnold, London
12. Scerrati M, Rossi GF (1984) The reliability of stereotactic biopsy. Acta Neurochir (Wien) [Suppl] 33: 201–205
13. Scerrati M, Fiorentino A, Fiorentino M, Pola P (1984) Stereotactic device for polar approaches in orthogonal systems. J Neurosurg 61: 1146–1147
14. Scerrati M, Rossi GF, Roselli R (1987) The spatial and morphological assessment of cerebral neuroectodermal tumours through stereotactic biopsy. Acta Neurochir (Wien) [Suppl] 39: 28–33
15. Scherer JH (1940) The form of growth in gliomas and their practical significance. Brain 63: 1–34
16. Schiffer D, Fabiani A (1975) I tumori cerebrali. Pensiero Scientifico, Roma
17. Walker MD (1983) Oncology of the nervous system. Martinus Nijhoff, Hingham
18. Zülch KJ (1979) Types histologiques des tumeurs du systeme nerveux central. Classification histologique internationale des tumeurs. OMS, Geneve
19. Zülch KJ (1980) Principles of the new World Health Organization (WHO) classification of brain tumours. Neuroradiology 19: 59–66
20. Zülch KJ (1986) Biological behaviour and grading (prognosis). In: Zülch KJ (ed) Brain tumours. Springer, Berlin Heidelberg New York, pp 27–40

Correspondence: Dr. Romeo Roselli, Istituto di Neurochirurgia, Università Cattolica S. Cuore, Largo A. Gemelli, 8, I-00168 Roma, Italy.

Acta Neurochirurgica, Suppl. 46, 82–85 (1989)
© by Springer-Verlag 1989

Stereotaxy of Third Ventricular Masses

E. Hitchcock and **B. G. Kenny**

University of Birmingham, Midland Centre for Neurosurgery and Neurology, Birmingham, U.K.

Summary

Posterior third ventricular masses are uncommon and posterior intraventricular masses are rare. Intraventricular and paraventricular lesions are difficult to differentiate in this region and present particular problems of diagnosis and treatment.

Uncertain radiology and frequently confusing histology make treatment planning difficult. The first requirement of histological confirmation of the speculative pathology is best achieved by stereotactic biopsy rather than craniotomy exposure and biopsy, the mortality and morbidity for each being compared.

The problem of CSF pathway obstruction can be dealt with by aspiration of the mass or CSF diversion; solid tumours frequently require a definitive shunting procedure while other lesions may not.

Keywords: Cerebral tumour; ventricle; stereotaxy.

Introduction

The third ventricle is frequently involved in a variety of masses, the majority of which are extraventricular in origin. True intraventricular third ventricular lesions constitute an interesting, but rare, sub-group, and include masses that "arise from the structures within the ventricular walls and have their point of attachment along the ventricular confines"[5]. Most commonly these are colloid cysts but include choroid plexus papillomas, ependymomas, epidermoids and dermoids, and subependymal gliomas. Non-neoplastic lesions may be congenital, infectious or vascular in origin[20].

The management of third ventricular masses presents particular problems. Radiological and biochemical evaluation may strongly suggest the underlying pathology in some lesions but it cannot give a definitive diagnosis and at worst is quite unhelpful[2, 6, 7, 11, 15, 18, 21]. All compartments of the third ventricular region are accessible by one or other of a variety of surgical approaches[5] but the end result of such exposure is frequently only a biopsy particularly in cases of intrinsic and invasive paraventricular lesions. The morbidity and mortality following such procedures is frequently considerable, and related principally to access; an example is memory disorder following fornix injury[19] and complex disorders of callosal function[8, 10, 12, 16]. Histological confirmation of diagnosis can be quickly and safely accomplished using CT guided stereotactic methods[1] and many lesions are then best managed by subsequent chemotherapy and irradiation[17]. There are many reports of stereotactic intervention in third ventricular tumours or cysts almost all being anteriorly situated or paraventricular in origin; pineal tumours are commonly and increasingly included. True posterior intraventricular third ventricular lesions are less common and we present examples to illustrate some of the particular problems and techniques available in the management of these lesions. Clinical presentation of masses within this region is characterized by disturbance in the visual pathways in the form of Paranaud's syndrome with or without accompanying hydrocephalus.

Materials and Method

The four cases reported presented to this unit between December and February 1988. They ranged in age from 14 to 70 years and are included because CT scanning revealed posterior intraventricular third ventricular lesions. Two were high attenuation lesions and in both of these cases vertebral angiography was performed to outrule vascular abnormalities.

Two patients had permanent CSF shunts inserted and a third had temporary emergency external ventricular drainage.

Technique and Instrumentation: In all cases the stereotactic frame was applied under local anaesthesia. The Hitchcock frame was used with a GEC 8800 scanner and appropriate targeting performed using the grid. Entry point selection was unnecessary except where ventriculoscopy was required in which case appropriate CT sagittal, coronal and oblique reconstruction allowed for choice of the best entry point and angle of trajectory (Fig. 1). All procedures were performed in a neurosurgical operating theatre under local anaes-

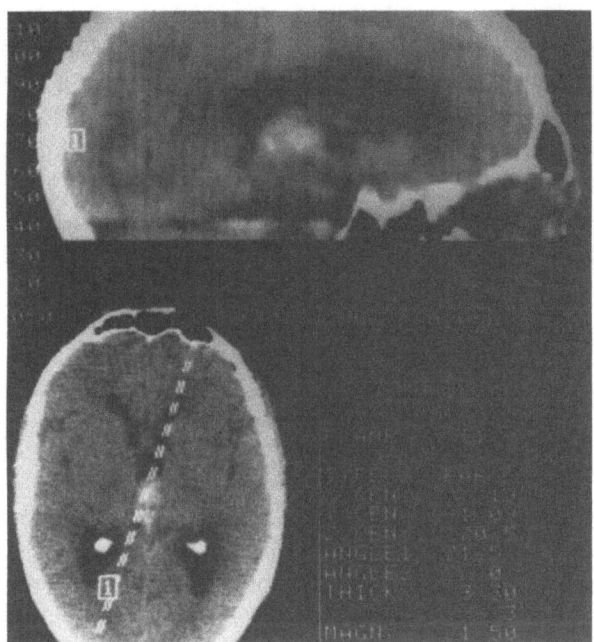

Fig. 1.

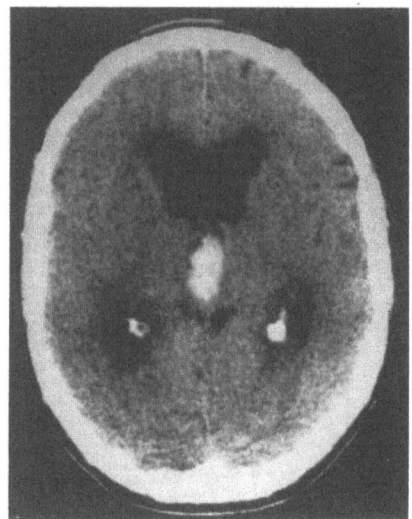

Fig. 3.

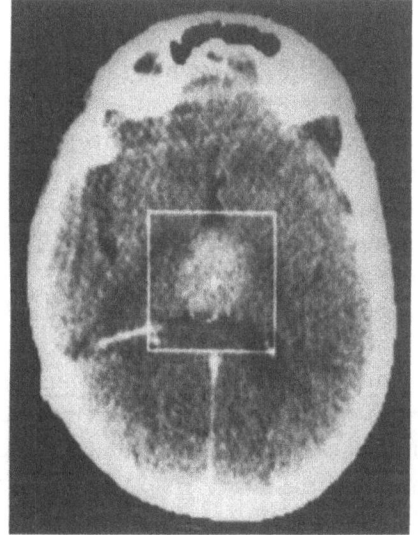

Fig. 2.

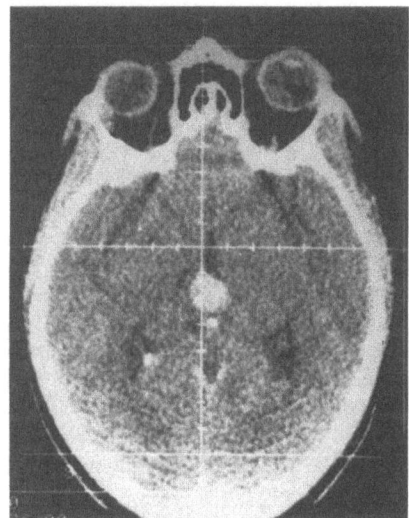

Fig. 4.

thesia without prior sedation. A Hitchcock stereotactic ventriculo-scope (Rimmer Bros.) was introduced through a standard burr hole in the right frontal region into the lateral ventricle and through the foramen of Monro at an angle pre-determined by CT targeting per-mitting identification of the lesion under direct vision, biopsy and irrigation. Stereotactic biopsy was performed using a double lumen biopsy cannula.

Case Histories

Case 1: E. D., 48-year-old woman presented with a six week history of progressive deterioration and impaired consciousness. She

was drowsy, disorientated with gross memory disorder and ataxia with generalized hypertonia. The CT scan is illustrated (Fig. 2). Her grave condition immediately improved after insertion of a VP shunt and subsequent stereotactic biopsy revealed a glioblastoma multi-forme. Stereotactic external beam irradiation was administered post-operatively and 15 months later she remains asymptomatic. The CT scan reveals residual thickening but no visible mass.

Case 2: R. M., 61-year-old male presented with acute frontal headache, nausea, vomiting and subsequent deteriorating conscious-ness. He was rousable with Parinauds syndrome and disorder of lateral gaze. CT scan illustrated (Fig. 3). Angiography revealed no vascular abnormality. Stereotactic ventriculoscopy showed the lesion to be a haematoma. Post-operatively his condition steadily improved and his eye signs resolved. Subsequent CT scanning both in the immediate and late postoperative period revealed disappearance of the haematoma and normal sized ventricles. Follow-up angiography revealed no evidence of AVM. The patient is asymptomatic.

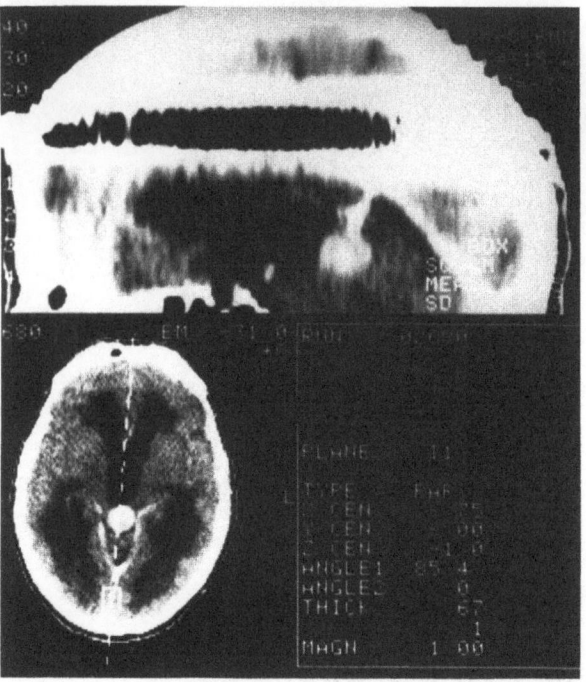

Fig. 5.

Case 3: M. P., a 14-year-old girl who had a 5 day history of headache and vomiting followed by acute collapse and coma. The pupils were unequal and fixed to light. CT scan illustrated (Fig. 4). Emergency external ventricular drainage produced immediate clinical improvement. Examination several hours later revealed Parinaud's syndrome. Angiography revealed no vascular abnormality. At CT guided stereotactic biopsy 10 ml of haematoma with several tissue fragments was aspirated. Histology revealed numerous abnormal blood vessels suggestive of vascular malformation or hamartoma. The ventricular drain was removed, the patient steadily improved and the Parinaud's syndrome became less marked. Subsequent CT scanning revealed resolution of the haematoma with no evidence of recurrence of the hydrocephalus. Repeat angiography revealed no abnormality.

Case 4: A. H., a 70-year-old man developed progressive confusion and disorientation. Chest X-ray revealed right lower lobe mass CT scan illustrated (Fig. 5) showed a posterior intraventricular lesion with secondary hydrocephalus. At CT guided ventriculoscopy a fleshy, frondular lesion was demonstrated which proved to be a metastatic adenocarcinoma. The external ventricular drain was left *in situ* and stereotactic external beam irradiation administered. He subsequently required permanent shunting. There was some improvement but he remained confused and disorientated.

Results

The procedure was well tolerated by all four patients and there were no operative complications. The quality of CT scanning in confused or agitated patients is considerably improved using the stereotactic frame as the frame is firmly anchored to the holding device on the CT table. CT targeting successfully located the lesion

and material was obtained for histological examination in three patients. In the fourth the lesion which proved to be an haematoma was successfully visualized with the ventriculoscope and flushed away. Both patients with tumours had externally delivered stereotactic focal irradiation in the immediate postoperative period. Two patients with tumours, required permanent shunting. Neither of the two patients with haematoma required permanent shunting and subsequent CT scanning has revealed no evidence of recurrent hydrocephalus.

Discussion

CT guided stereotactic biopsy of lesions of the third ventricular region have been well described in the past[1]. Aspiration of colloid cyst with resolution of abnormality and avoidance of permanent shunting has also been described[4, 14]. A variety of lesions present in this region; Apuzzo reported the management of 42 "third ventricular" lesions of which 8 were anterior region, 8 posterior region, 23 paraventricular and only 3 were true intraventricular masses of which one was a colloid cyst. Our cases illustrate the varied nature of true posterior intraventricular lesions. The development of stereotactically guided instruments extends the surgical options and the actual procedure may in itself be therapeutic (cases 2 and 3) so that not all patients require permanent CSF pathway shunting. Emergency external drainage allows clinical improvement to facilitate stereotaxic planning and post-operative scanning will reveal the need for permanent shunting.

Preoperative angiography may be necessary in many cases but ventriculoscopy by allowing direct lesion visualization often reduces this need. Further instruments have been described by other authors which can be used with the ventriculoscope system including cutting scissors, biopsy forceps, lasers and diagnostic ultrasonography[3, 9]. The advent of stereotactically delivered radiosurgery adds a further refinement to the management of these lesions[13].

Acknowledgements

We would like to thank the CT Department, Miss Sue Redfern, Mr. William Mitchell and the Department of Medical Illustration.

References

1. Apuzzo Michael LJ *et al* (1984) Computed tomographic guidance stereotaxis in the management of lesions of the third ventricular region. Neurosurgery 15: 502–508
2. Arita N, Biroh S, Ushio Y *et al* (1983) Primary pineal endodermal sinus tumour with elevated serum and CSF alfafoetoprotein levels. J Neurosurg 53: 244–248

3. Auer LM, Holzer P, Ascher PW, Heppner F (1988) Endoscopic neurosurgery. Acta Neurochir (Wien) 90: 1–14
4. Bosch DA, Rahn T, Backlund EO (1978) Treatment of colloid cysts of the third ventricle by stereotactic aspiration. Surg Neurol 9: 15–18
5. Carmel PW (1985) Tumours of the third ventricle. Acta Neurochir (Wien) 75: 136–146
6. Chang CG, Kageyama N, Kobayashi T, Yoshida J, Negoro N (1981) Pineal tumours: Clinical diagnosis, with special emphasis on the significance of pineal calcification. Neurosurgery 8: 656–668
7. Edwards MSB *et al* (1988) Pineal region tumours in children. J Neurosurg 68: 689–697
8. Gazzaniga MS, Risse GL, Springer SP, Clark E, Wilson DH (1975) Psychological and neurologic consequences of partial and complete cerebral commissurotomy. Neurology 25: 10–15
9. Heikkinen ER, Heikinnen MI (1987) New diagnostic and therapeutic tools in stereotaxy. Appl Neurophysiol 50: 136–142
10. Heilman JF, Zauaoui A, Renier D, Pier-Kahn A (1979) A new surgical approach to the third ventricle with interruption of the thalamostriate vein. Acta Neurochir (Wien) 47: 135–147
11. Hill S, Martin EC, Ellison EC, Hunt WE (1980) Carcinoembryonic antigen in cerebrospinal fluid of adult brain tumour patients. J Neurosurg 53: 627–632
12. Jeeves MA, Simpson DA, Geffen G (1979) Functional consequences of the transcallosal removal of intraventricular tumours. J Neurol Neurosurg Psychiatry 42: 134–142
13. Leksell DG (1987) Stereotactic radiosurgery, present status and future trends. Neurol Res 9: 60–68
14. Musolino A *et al* (1987) Stereotactic aspiration of colloid cysts of the third ventricle. Preliminary report. Appl Neurophysiol 50: 210–217
15. Pomarede R *et al* (1982) Endocrine aspects and tumoural markers in intracranial germinoma: An attempt to delineate the diagnosis procedure in 14 patients. J Paediatr 101: 374–378
16. Rhotom AL, Yamamoto I, Peace DA (1981) Microsurgery of the third ventricle: Part 2: Operative approaches. Neurosurgery 8: 357–373
17. Sung D, Harisiadis L, Chang CH (1978) Midline pineal tumours and suprasellar germinomas: highly curable by irradiation. Radiology 128: 745–751
18. Suzukin Y, Tanaka R (1980) Carcinoembryonic antigen in patients with intracranial tumours. J Neurosurg 53: 355–360
19. Sweel WH, Talland GA, Ervin FR (1959) Loss of recent memory following section of fornix. Trans Am Neurol Assoc 84: 76–82
20. Vaquero J *et al* (1980) Cavernous angiomas of the pineal region. J Neurosurg 53: 833–835
21. Zimmerman RA, Bilaniok LR, Wood JH, Bruce DA, Schut L (1980) Computed tomography of pineal, parapineal and histologically related tumours. Radiology 137: 669–677

Correspondence: Professor E. Hitchcock, University of Birmingham, Midland Centre for Neurosurgery and Neurology, Holly Lane, Smethwick, Birmingham, B67 7JX, U.K.

Acta Neurochirurgica, Suppl. 46, 86–89 (1989)
© by Springer-Verlag 1989

Brain Stem Expanding Lesions: Stereotactic Diagnosis and Therapeutical Approach

F. Giunta, G. Grasso, G. Marini, and F. Zorzi[1]

Neurosurgical Department of the University of Brescia, [1] Service of Pathology, Spedali Civili, Brescia, Italy

Summary

In most cases of brain stem expansive lesion a surgical approach is possible, but in each patient it must be evaluated if the surgical risk is proportional to the therapeutic result. Sometime surgery is limited to a biopsy sample, particularly in malignant lesions. We started stereotactic serial biopsy sampling in all CT or NMR intra-axial brain stem expansive lesions as a preliminary diagnostic procedure. The aim is to look for benign well delimited lesions that we consider for surgical removal or to drain haematomas and abscesses.

35 patients with brain stem expansive lesions were submitted to 47 surgical procedures: 35 stereotactic biopsies (one each patient) and, among them, 12 were major surgical procedures (with craniotomy) for microsurgical removal of the expansive lesions. 15 patients were in paediatric age.

Suboccipital transcerebellar approach was performed in 25 mesencephalic, pontine, bulbar expansive lesions and frontal approach was limited to 10 thalamo-mesencephalic lesions. There was no mortality. Two patients were stereotactically drained and definitively cured.

Keywords: Brain stem; brain tumour; stereotactic biopsy.

Introduction

Surgery of the brain stem expanding lesions is considered a challenge by neurosurgeons mainly because 1) difficulties in approach[3], 2) fear of impairing vital cerebral structures[1, 27, 33, 35] and 3) consciousness that in many instances the surgery does not cure the disease and in many cases only biopsy is possible.

Stereotactic serial biopsy is a method to obtain a pathological diagnosis from any cerebral lesion everywhere it is sited into the brain. If an angiographic study is done and a vascular malformation excluded, the stereotactic surgical procedure is simple and reliable. It provides useful seriated specimens along the CT morphology of the lesion and in a high percentage of cases (from 88 to 94%) in brain stem or supratentorial lesions respectively an histological diagnosis is achieved.

We have to keep in mind, when an expanding lesion is seen at CT scan: 1) that not all the expansive lesions are neoplasms, 2) that sometime the lesions are surgically approachable (well delimited slow growing neoplasms) and 3) that cystic lesions can be drained. That is true particularly for brain stem lesions mainly because of the risk of open surgery.

Brain stem open surgery[3, 9, 11, 17] or stereotactic drainage[2, 4, 5, 7, 10, 12] are able to cure selected patients.

A therapeutic problem arises when a slow growing tumour is diagnosed. In these cases radiotherapy or chemotherapy are ineffective. But in selected cases, if tumours are well delimited, surgical removal can be attempted and the outcome may be modified.

Material and Method

Whole data are summarized in Table 1.

33 patients with brain stem expanding lesions were referred to the Neurosurgical Department of the University of Brescia from July 1983 to May 1988 for stereotactic serial biopsies.

14 were males and 19 females. The age ranged from 2 to 69 and, among them, 18 were under 15 years.

Before the stereotactic procedure, 4 patients had a CSF shunt, 2 patients were previously submitted to radiotherapy without histological diagnosis and one patient underwent a stereotactic bioptic procedure in another centre, without a pathological diagnosis, for insufficient tissue sampling.

The overall clinical history had an average of 9 months, but many patients suffered early symptoms only one month before the stereotactic biopsy. The most frequent symptoms were: intracranial hypertension syndrome, diplopia, motor disturbances, V-VII-VIII cranial nerves impairment. The performance status was about 60 of the Karnofsky scale.

All patients had one or more CT scans, 12 had MRI studies and 18 had vertebral angiography. The tumour bulk was in the mesencephalo-ponto-bulbar region, but in 6 patients the fourth ventricle was occuped, in 4 patients a tumour infiltrated the cerebellar peduncle, and in one case it extended to the medulla at C2 level. The

Table 1. *Summary of Clinical, Stereotactic and Treatment Data of 33 Patients Bearing Brain Stem Expanding Lesions.* Neurosurgical Department, University of Brescia, 1983–1988

[Large rotated multi-column data table with columns: PAT (N. 33), SEX, AGE, PREV-TREAT (STE, SH, RxT), SYMPTOMS (ICH, CO, 346, 578, PA, MO, SE, GT, TE, CC), KAR, SITE, ANAESTHESIA (LO, NLA, GE), APPROACH (CV, TCE, TE, SPL, AS), POST-OP-CT (N, NF, AIP), SIDE-EFFECTS (VO, WO, UC, AM, CU), HISTOLOGY, POST-STEREOTACTIC TREATMENT (BIO, SH, DR, SU, CHT, RXT, CU, NT), 2nd HISTOLOGY. Individual patient data rows, largely illegible in detail.]

PAT	SEX	AGE	SITE	HISTOLOGY	2nd HISTOLOGY
SE	F	10	90 mese+ic	pilocitic a.	pilocitic a.
NF	M	49	20 mese+ic	pilocitic a.	pilocitic a.
PA	F	8	30 mese+ic	pilocitic a.	protoplasmatic a.
VP	F	59	60 mese+ic	protoplasmatic a.	ependymoma
AL	F	59	40 mese	papilloma	
ML	F	7	70 mese	fibrillar a.	
MO	F	7	40 mese	anaplastic a.	
PM	F	68	50 mese	anaplastic a.	anaplastic a.
PM	F	25	90 mese+pons	not diagnosed	
LG	F	68	70 mese+pons	melanoma	
MC	M	27	59 mese+pons	anaplastic a.	
RF	F	19	50 mese+pons	oligo-astrocytoma	
TR	M	35	50 mese+pons+ce	anaplastic a.	
AM	F	5	80 pons	fibrillar a.	
CA	M	5	70 pons	fibrillar a.	
DM	M	17	30 pons	glioblastoma	
VS	M		40 pons	pilocitic a.	
AG	M	15	90 pons+4v	fibrillar a.	
ML	F	17	60 pons+4v	ependymal cyst	ependymal cyst
SE	M	14	70 pons+4v	anaplastic a.	
ZM	M	15	80 pons+4v	pilocitic a.	
BC	M	4	70 pons+bulb	ependymoma	ependymoblastoma
GO	F	14	40 pons+bulb	hematoma(mav)	
GL	F	12	55 pons+bulb	pilocitic a.	
SL	F	3	50 pons+bulb	pilocitic a.	
DS	M	3	90 pons+bulb	fibrillar a.	
CL	M	3	50 pons+bulb+4v	ependymoma	
BI	M	26	59 pons+bulb+ce	fibrillar a.	
PE	F	64	80 pons+bulb+ce	not diagnosed	
BE	F	52	40 bulb(clivus)	meningioma	meningioma
MG	M	2	80 bulb+4v	not diagnosed	anaplastic a.
RL	M	69	50 bulb+ce	metastasis	
BF	M	22	70 bulb+med	not diagnosed	pilocitic a.

PREV-TREAT=previous treatment: STE=stereotactic biopsy, SH=c.s.f. shunt, RxT=radiotherapy.

SYMPTOMS: ICH=intracr.hypertension,CO=consciousness imp,346=oculomotion imp,578=cranial nerves imp,PA=Parinaud,MO=motor imp,SE=sensory imp,GT=gait imp,TE=tetraparesis,CC=cerebellar incoordination.

KAR=Karnofsky status scale. SITE: mese=mesencephalon,ic=internal capsula,ce=cerebellum,bulb=bulbar,4v=fourth ventricle,med=medulla. ANAESTHESIA: LO=local,NLA=neuroleptoanalgesia,GE=general.

APPROACH: CV=convexity,TCE=transcerebellar. TG=targets. SPL=samples. AS=cysts aspirations. POST-OP-CT: N.=numbers,NF=nul findings,AIP=air bubbles.

SIDE-EFFECTS: VO=vomiting,WO=transitory neurological worsening,UO=unchanged,AM=ameliorated,CU=cured.

POST-STEREOTACTIC TREATMENT: BIO=stereotactic biopsy,SH=c.s.f.shunt,DR=surgical drainage,SU=surgical removal,CHT=chemotherapy,RxT=radiotherapy,CU=curietherapy (Ir192),NT=not treated.

major expansion was: in 13 cases in the right side, in 14 in the left and in 6 in the midline.

Because of the high risk with open surgery due to: 1) the extension of the expansive lesion, 2) the young age of many patients (18 cases) or the advanced age (3 cases elder than 65), 3) the advanced illness (8 cases had 50, 5 cases had 40, 2 cases had 30 and 20 each of Karnofsky index), the indication to stereotactic biopsy was posed.

18 stereotactic procedures have been carried out in local anaesthesia, 6 cases in neuroleptoanalgesia while younger children underwent general anaesthesia.

In 26 patients the lesion was approached through a suboccipital transcerebellar route with the patient in the sitting position. In 7 patients, with mesencephalic lesions, a coronal approach was preferred.

Serial biopsy samples, along the choosen trajectory, were taken into the 35 calculated targets (in 2 patients two targets were calculated) and a total of 96 samples were studied with cytological and histological techniques. In 5 cases a cystic fluid was aspirated in one case the wall membrane was taken out with a spiral needle (ependymal cyst).

Results

Of 35 calculated targets in 33 patients a number of 96 samples were examined. Cytological and histological studies showed: 9 malignant tumours (5 anaplastic astrocytoma, 1 oligo-astrocytoma with anaplasia, 1 glioblastoma, 1 metastasis, 1 primitive melanoma), 18 slow growing tumours (7 pilocytic astrocytoma, 6 fibrillar astrocytoma, 1 protoplasmatic astrocytoma, 2 ependymoma, 1 choroidal plexus papilloma and 1 meningioma), and 2 non-neoplastic expanding lesions (1 ependymal cyst and 1 haematoma due to a cryptic angioma).

In 4 cases the pathological examination showed normal tissue: in 1 patient stereotactic biopsy was repeated and anaplastic tissue was found, 2 patients with well delimited expanding lesions underwent a suboccipital craniotomy (a bulbar pilocytic cystic astrocytoma, respectively an intra-fourth ventricle fibrillary astrocytoma were found), 1 patient (a 64-year-old woman with a diffuse ponto-bulbar hypodensity at CT scan) refused to repeat the stereotactic biopsy, but the follow-up suggests a slow growing neoplasm.

A diagnosis with stereotactic serial biopsies was obtained in 29 of 33 patients (88%).

After stereotactic surgery the clinical condition remained unchanged in 25 patients. In 4 patients a transitory worsening was observed. On the contrary 3 patients with a cystic expanding lesions ameliorated because of the reduction of the cyst volume. Vomiting was present in 2 patients. One child was cured by stereotactic aspiration of an haematoma due to a cryptic arteriovenous malformation. We had no mortality.

After stereotactic biopsy the treatment was suboc-

cipital craniotomy in 10 patients: 3 had a cystic lesion that was opened into the CSF space, one of these had to be reopened two months later because the cystotomy was closed, and in 7 patients a microsurgical tumour removal was performed; one patient was operated twice for a recurrent ependymoma. Radiotherapy was done in 12 and chemotherapy in 3 patients. Stereotactic biopsy was repeated in 1 and curietherapy with ^{192}I was applied. In 5 patients there was no further therapeutic approach.

Discussion

The main indication for stereotactic serial biopsy of CT visualized intracranial lesions is to provide a pathological diagnosis as accurate as possible for a better therapeutic approach. This possibilities are true mainly in brain stem expanding lesions because open surgical approach is hard and often inadequate.

If we know the pathology we can better plan the therapy. We suggest:

1) Anaplastic tumours are not cured by any known therapeutic approach: surgery, radiotherapy, chemotherapy or immunotherapy are today all palliative. Although radiotherapy is proven to give a longer survival than other therapies.

2) Slow growing tumours are suitable for microsurgical removal only when they are well delimited from surrounding structures or are cystic. In this case they should be opened into the CSF space and the tumour nodule, if present, removed. If the cyst remains patent we observed a remission of symptoms and no progression of the disease was observed.

3) When the expanding lesions is not a neoplasm, surgical removal is mandatory. Sometimes stereotactic drainage is sufficient to cure the disease (haematomas, abscesses).

References

1. Albright AL, Sclabassi RJ (1985) Use of the Cavitron ultrasonic surgical aspirator and evoked potentials for the treatment of thalamic and brain stem tumours in children. Neurosurgery 17: 564–568
2. Apuzzo MJ, Sabshin JK (1983) Computed tomographic guidance stereotaxis in the management of intracranial mass lesions. Neurosurgery 12: 277–285
3. Baghai P, Vries JK, Bechtel PC (1982) Retromastoid approach for biopsy of brain stem tumours. Neurosurgery 10: 574–579
4. Broggi G, Franzini A, Migliavacca F, Allegranza A (1983) Stereotactic biopsy of deep brain tumours in infancy and childhood. Child's Brain 10: 92–98
5. Broggi G, Franzini A, Giorgi C (1983) The value of stereotactic biopsy in management of brain stem lesions. Ital J Neurol Sci [Suppl] 21: 51–56

6. Coffey RJ, Lunsford LD (1985) Stereotactic surgery for mass lesions of the midbrain and pons. Neurosurgery 17: 12–18

7. Franzini A, Ferraresi S, Giorgi C, Allegranza A, Broggi G (1987) La biopsia stereotassica nel trattamento delle lesioni espansive del tronco cerebrale. In: B Cucciniello Attualita' in Neurochirurgia 1987: 23–27

8. Halperin EC (1985) Paediatric brain stem tumours: Patterns of treatment failure and their implications for radiotherapy. Int J Radiat Oncol Biol Phys 11: 1293–1298

9. Heppner F, Oberbauer RW, Ascher PW (1985) Direct surgical attack on pontine and rhombencephalic lesions. Acta Neurochir (Wien) [Suppl] 35: 123–125

10. Mathisen JR, Giunta F, Marini G, Backlund E-O (1987) Transcerebellar biopsy in the posterior fossa: 12 years experience. Surg Neurol 27: 297–299

11. Murphy MG (1972) Successful evacuation of acute pontine haematoma: case report. J Neurosurg 37: 224–225

12. Niizuma H, Suzuki J (1987) Computed tomography-guided stereotactic aspiration of posterior fossa haematomas: a supine lateral retromastoid approach. Neurosurgery 21: 422–427

13. Ravetto F, Di Cagno L, Bosco M, Madon E, Gajno TM (1979) I tumori infiltranti del tronco encefalico in eta' evolutiva. Min Ped 31: 95–110

14. Russell B, Rengachary SS, McGregor D (1986) Primary pontine haematoma presenting as a cerebellopontine angle mass. Neurosurgery 19: 129–133

15. Soffer D, Sahar A (1982) Cystic glioma of the brain stem with prolonged survival. Neurosurgery 10: 499–502

16. Stroink AR, Hoffman HJ, Hendrick EB, Humphreys RB (1986) Diagnosis and management of paediatric brain stem gliomas. J Neurosurg 65: 745–750

17. Van Gilder JC, Allen WE, Lesser RA (1974) A pontine abscess: survival following surgical drainage: case report. J Neurosurg 40: 386–390

18. Villani R, Gaini SM, Tomei G (1975) Follow-up of brain stem tumours in children. Childs Brain 1: 126–135

Correspondence: F. Giunta, M.D., Neurochirurgia, Spedali Civili, I-25100 Brescia, Italy.

Acta Neurochirurgica, Suppl. 46, 90–93 (1989)
© by Springer-Verlag 1989

Tumour Recurrence vs Radionecrosis:
an Indication for Multitrajectory Serial Stereotactic Biopsies

L. Zamorano, D. Katanick, M. Dujovny, D. Yakar, G. Malik, and **J. I. Ausman**

Henry Ford Neurosurgical Institute, Department of Neurological Surgery, Henry Ford Hospital Division, Detroit, MI, U.S.A.

Summary

External RT has been proved to be an important adjuvant to surgery in the treatment of malignant glioma. It has also been demonstrated, that its effect on survival is dose-dependent, although accompanied by a higher morbidity. Intents to localize the field of high dose RT to the tumour area have been performed with the aim to spare damage of the normal brain tissue. Between August 1983 to December 1987, 40 patients with malignant astrocytoma (16 GM, 24 AA) underwent high dose localized hyperfractionated external RT after surgical resection. Patients received 57.6 Gy to the tumour and oedema area associated with a boost localized to the tumour of 7.4, 14.4 or 24 Gy. In the follow-up, 16 patients died with evidence of increase in size of lesion diagnosed by CT/MRI. Since July 1987, 12 patients with recurrence or increase on size of CT/MRI lesion have undergone multitrajectory serial stereotactic biopsies. From the biopsies 8 patients were histologically diagnosed was compatible with radionecrosis. From the 4 recurrences, 2 patients were treated with ^{125}I implants and 1 with new resection. Patients with radionecrosis were treated with corticoides and diuretics, obtaining partial or complete remission of symptoms and decrease in size of CT lesion. Undoubtly, Multiplanar/3 D multitrajectory serial stereotactic biopsies play a major role in the follow-up of these patients, and accurate diagnosis need to be established for further treatment therapy. The question remains if these localized boost should be replaced by 3 D Multiplanar stereotactic interstitial radiotherapy boost after surgery and conventional radiotherapy.

Keywords: Radionecrosis; malignant gliomas; stereotactic biopsies.

Introduction

External radiotherapy has been proven to be an important adjuvant to surgery in the treatment of malignant gliomas[1, 3, 6]. It has been demonstrated that its effect on survival is dose-dependent, although accompanied by a higher morbidity[2]. Attempts to localize the field of high dose external radiation to the tumour area have been made with the aim to avoid damage to normal brain tissue. Localization is generally performed using the conventional two-film technique. At our institution forty patients with malignant astrocytoma underwent surgical resection followed by high dose localized external radiation therapy (range 6,480–8,160 cGy) with concomitant chemotherapy with BCNU. Survival analysis and evaluation of treatment is presented and the role of multitrajectories serial image-guided stereotactic biopsies at the time of progression is discussed.

Material and Methods

From August 1983 until December 1987, forty patients were treated with surgical resection followed by high dose localized external radiation therapy (range 6,480–8,160 cGy) and concomitant chemotherapy with BCNU after. Pathological diagnosis revealed a malignant glioma in all cases: anaplastic astrocytoma on 17 cases, glioblastoma multiforme on 8, 7 mixed oligo-astrocytoma, 8 mixed glioblastoma and sarcoma. Two patients received high dose localized radiation (6,500 and 7,000 cGy) with a regime of 200 cGy per day. Thirty-eight patients had hyperfractionated regime of 120 cGy twice a day. They received the first 5,760 cGy to the area of the tumour volume plus oedema plus 2 cm margin defined by preoperative CT. Subsequent additional doses of 740, 1,440, 2,400 cGy (randomized) were delivered to a reduced field to include the tumour plus 2.5 cm margin after consideration of the preoperative contrast CT scan. Treatment fields were modified with the use of cast blocks in order to confine the treatment volume and spare the damage of normal brain tissue.

Twenty-nine patients had documented progression:

group A-(August 1983–June 1986) 17 cases, had no further biopsy or reoperation, and

group B-(July 1986–December 1987) 12 cases were evaluated with multitrajectories serial stereotactic biopsies. This constitutes a non-randomized group and reflects the time when we started the performance of Image-guided Stereotaxis at our Institution.

Multitrajectories Serial Stereotactic Biopsies Methodology

Double dose contrast, 1.5 mm thick slice high resolution CT scan of the lesion with placed stereotactic frame.

At the CT console, size, shape, and main axis of target volume is defined by using 2D Multiplanar approach. Interactive location of targets and trajectories was performed to optimize the histological sampling. Image-defined coordinates and angles were transposed into stereotactic space by a computer program that define arch angles and depths.

Entry point at the skin level is marked after setting of arch parameters. Under local anaesthesia, a burr hole is performed at the decided entry site. Serial biopsies at 5 mm intervals are performed, starting 1–2 cm above the image-defined border and finishing 1–2 cm beyond the margins (Fig. 1). Through the same burr hole another 2 or 3 trajectories can be performed following the same technique.

Eight patients had histological diagnosis of radionecrosis and 4 of "real recurrences". 2 of 4 patients with malignant recurrent tumours were treated with high-activity ^{125}I interstitial radiotherapy.

Statistical Analysis

Survival analysis was performed using the Kaplan-Meier method. At date of analysis (December 1987), 23 patients were dead and 17 were alive. Survival was considered from the date of original brain surgery. The log rank test was used to perform various subgroups survival comparisons, where p-values less that 0.05 indicate a statistically significant difference. Cox regressional analysis was used in a multivariate settings, stepwise proportional hazards survival analysis.

Results

The overall median survival of the series was 622 days (Fig. 2). Median survival time was directly related to age, 21 patients younger than 50 years had median survival of 724 days compared to 19 patients older than 51 years with median survival of 389 days (p < 0.006). Intital Karnofsky and neurological grade were not statistically significant predictors of survival. Follow-up

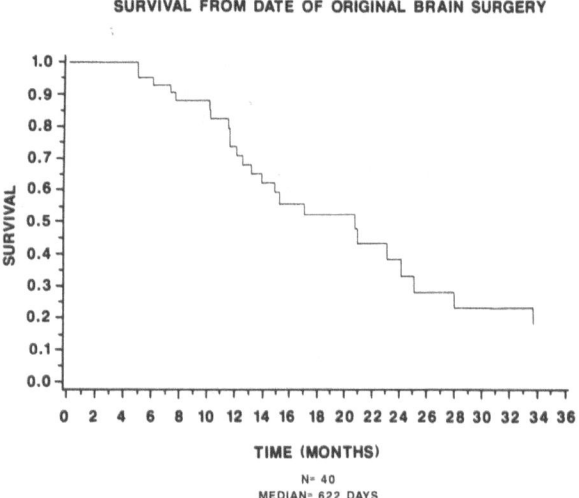

Fig. 2. Overall survival

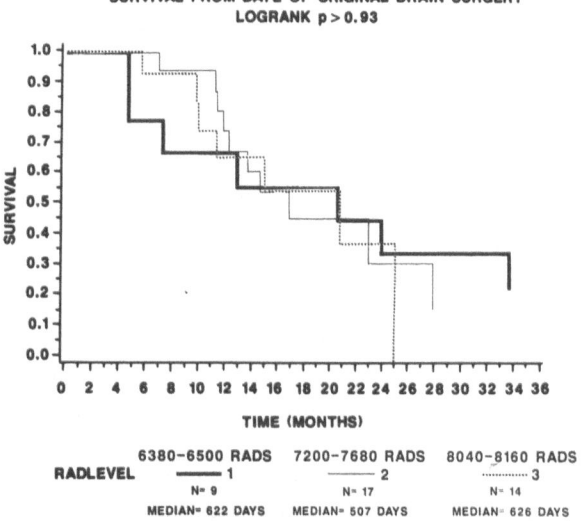

Fig. 3. Survival according total radiation dose to the tumour volume

Karnofsky (p < 0.009) and follow-up neurological grade (p < 0.04) were important determinants of survival. Type of macroscopic resection (p < 0.004) indicated strong correlation, total resection having longer survival (median = 753 days) than subtotal resection (median = 389 days). Symptoms duration and histology were indicated as possible predictors of survival with p-values of 0.05. Total dose of radiation did not show a significant improvement in survival (Fig. 3). The stepwise proportional hazards survival analysis results indicate that 1. follow-up Karnofsky, 2. stage, and 3. histology are, in that order of influence, the statistically significant prognostic predictors of survival from date of original surgery at the 0.05 significance level.

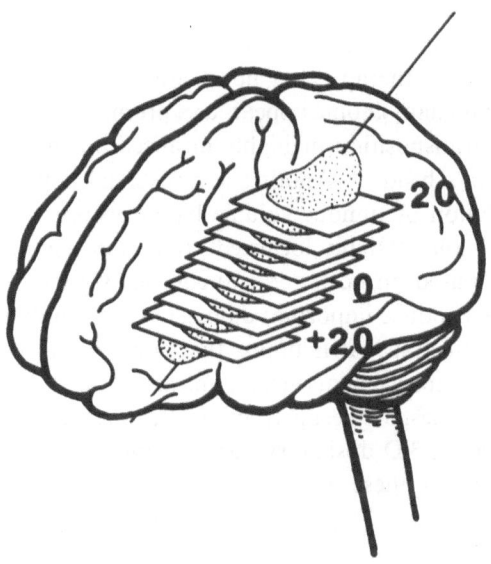

Fig. 1. Schematic representation of one trajectory of serial stereotactic biopsies

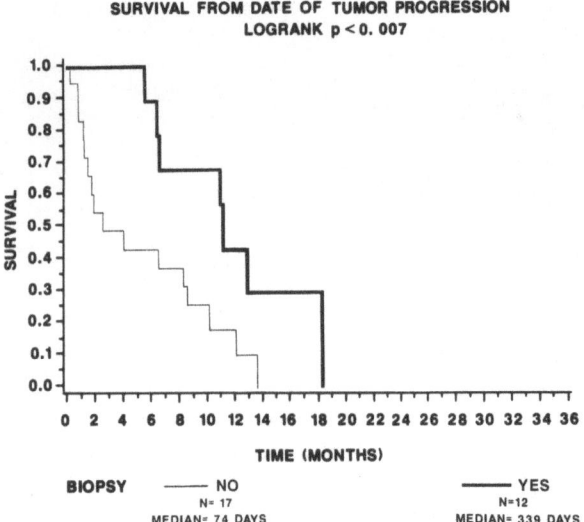

Fig. 4. Survival comparing patients with and without multitrajectories serial stereotactic biopsies at the time of progression

Considering the subgroup of 29 patients who had progression of disease, we compared the 12 patients with multitrajectories serial stereotactic biopsies with the 17 without biopsy, considering survival from time of tumour progression: patients with biopsies (7 of 12 died) had a median survival of 339 days from date of progression. Patients without biopsies (16 of 17 died) had a median survival of 74 days from date of documented progression. The resulting log rank test indicated a high statistical significance (p < 0.007).

Considering the 12 patients who had stereotactic biopsies, the 8 patients with histological picture of radionecrosis were compared to the 4 patients with "real tumour recurrence" regarding survival from the date of tumour progression: of the radionecrosis patients 5 of 8 died and median survival from date of tumour progression was 199 days; of the recurrence patients 3 of 4 died and median survival from date of tumour progression was 448 days (p > 0.08) (Fig. 4). Two of the 4 patients with recurrence underwent stereotactic 125I interstitial radiotherapy, one had a second craniotomy and one patient was not treated because of his poor clinical conditions. The log rank test p > 0.08 indicates that a statistically significant survival difference between the 8 radionecrosis patients and the 4 recurrence patients could be detected. However, the test power was very limited due to the small sample size.

Discussion

Postoperative radiation has been proven to increase the length of survival and median survival times in patients with malignant gliomas. Effectiveness is limited by the tolerance of the CNS to irradiation. Effects on survival are dose-dependent and median survival for patients who received 50 Gy has been reported 28 weeks, compared with 36 weeks for 55.5 Gy and 42 weeks for 60 Gy[5]. Intents to increase the total dose as high as 80 Gy have been performed: it was shown an increase in median survival related to increase of total dose, but with increase in morbidity[1, 2]. Intents have been performed to localize the fields of external radiation to the tumour area. This study represents such efforts. Our median survival of 622 days represents an improvement in survival compared with historical groups[1, 3, 5, 6]. Statistically significant predictors of survival were: age, T-stage (AJCC classification), type of resection, postoperative results, follow-up neurological grade, and follow-up Karnofsky rating.

Multitrajectories serial histological mapping allowed accurate and safe sampling to characterize tissue at time of progression. At the time of progression (increase on size of the lesion), multitrajectories serial image-guided stereotactic biopsies allowed an accurate mapping and histological evaluation of the lesions. Therapeutic consideration are derived of this histological diagnosis. Although high dose localized external radiotherapy increases survival in patients with malignant astrocytomas compared with historical groups radionecrosis short median survival (199 days) compared to "real recurrence" median survival (448 days) from date of progression implies that further efforts need to be done in order to localize high dose radiation treatments. Contributors to improvements in localization will probably be 3-D algorithms applied to Image display, treatment plannings and dosimetry. Stereotactic interstitial radiotherapy as a boost after initial diagnosis of malignant glioma appears as a promisory technique for transposition of highly localized dosimetry into patient's head[4, 7]. Another controversial issue is the calculation of tumour volume based on preoperative CT scans. "Highly localized" radiation therapy should be based on postoperative contrast CT otherwise it will produce important damage in surrounding normal brain tissue. It is hoped that combination of factors, like improvements in anatomical and metabolic image modalities, 3 D reconstructions, accuracy of surgical resection, 3 D dosimetry, use of radiosensitizers, etc will improve these results.

References

1. Salazar O, Rubin P, McDonald J, Feldstein M (1976) High dose radiation therapy in the treatment of glioblastoma multiforme: a preliminary report. Int J Radiat Oncol Biol Phys 1: 717–727

2. Sheline GE, Wara WM, Smith V (1980) Therapeutic irradiation and brain injury. Int J Radiat Oncol Biol Phys 6: 1215–1228

3. Leibel SA, Sheline GE (1987) Radiation therapy for neoplasms of the brain. J Neurosurg 66: 1–22

4. Szikla G, Schlienger M, Blond S, Daumas-Duport C, Missir D, Miyahara S, Musolino A, Schaub C (1984) Interstitial and combined interstitial and external irradiation of supratentorial gliomas. Results in 61 cases treated 1973–1981. Acta Neurochir (Wien) [Suppl] 33: 355–362

5. Walker M, Strike T, Sheline G (1979) An analysis of dose-effect relationship in the radiotherapy of malignant gliomas. Int J Radiat Oncol Biol Phys 5: 1725–1731

6. Wara WM (1985) Radiation therapy for brain tumours. Cancer 55: 2291–2295

7. Zamorano L, Dujovny M, Malik G, Yakar D, Mehta B (1987) Multiplanar CT guided stereotaxis and I-125 interstitial radiotherapy. Appl Neurophysiol 50: 281–286

Correspondence: Lucia Zamorano, M.D., Ph.D., Department of Neurological Surgery, Henry Ford Hospital, 2799 West Grand Boulevard, Detroit, MI 48202, U.S.A.

Acta Neurochirurgica, Suppl. 46, 94–96 (1989)
© by Springer-Verlag 1989

Comments on Brachycurie Therapy of Cerebral Tumours

M. Scerrati, R. Roselli, M. Iacoangeli, P. Montemaggi[1]**, N. Cellini**[1]**, R. Falcinelli**[1], and **G. F. Rossi**

Institute of Neurosurgery and [1] Institute of Radiology, Catholic University, Rome, Italy

Summary

Between 1980 and 1987 thirty patients harbouring cerebral neuro-epithelial tumours have been treated with stereotactic brachycurie therapy (18 males, 12 females), either alone (n = 16) or combined with surgery (n = 7) and/or external radiotherapy (n = 10). There were 25 slowly growing tumours (grade I n = 1; grade II n = 24). The remaining 5 were malignant tumours (grade III n = 3; grade IV n = 2). The radioactive sources utilized were ^{192}Ir in 26 cases and ^{125}I in 4. Twenty-eight patients underwent permanent implantation, the other two received temporary irradiation with removable after-loaded catheters. Target volume was less than 15 cm^3 in 6 cases, between 16–60 cm^3 in 17 and more than 60 cm^3 in 7. Tumour dose at the periphery of the target volume was: 70–100 Gy in 19 and 100–130 Gy in 9 of the cases treated with permanent implantation; the patients irradiated with removable implants received 40–60 Gy in 5–7 days. General follow-up ranged between 0.3 and 6.9 years (mean = 2.5 years). The results are analyzed with reference to the following aspects: 1) natural history of the disease; 2) modalities and goal of the treatment; 3) place of brachy therapy as sole treatment and combined with the other available therapeutical means.

Keywords: Brain tumours; stereotaxy; brachycurie therapy.

Introduction

Brachycurie therapy (BCT) permits, as is well known, the delivery of high radiation doses to an exactly defined tumour volume in respect of the tolerance threshold of the surrounding healthy tissue[1–7, 10]. The interest in this procedure in the treatment of brain tumours has increased in recent years by the advances of stereotactic surgery combined with the progress of the modern imaging techniques in neuroradiology.

Material and Method

Patients. From 1980 to 1987, 30 patients (mean age 34 ± 14.3 years, 18 males and 12 females) with cerebral neuroepithelial tumours were treated with stereotactic brachycurie therapy. Twenty-five patients harboured slow growing tumours (1 pilocytic astrocytoma, 15 fibrillary astrocytomas, 6 oligodendrogliomas, 3 oligoastrocytomas), 5 patients malignant tumours (2 anaplastic astrocytomas, 1 ana-

plastic oligodendroglioma, 2 glioblastomas). The tumour volume to be treated was assessed by integration of neuroradiological data with the results of previous stereotactic biopsy[5, 7–9]: it was less than 15 cm^3 in 6 cases, between 16 and 60 cm^3 in 17 and more than 60 cc in 7.

Method. Stereotactic BCT was carried out as the sole treatment in 15 patients: it was combined with surgery in 5 cases (2 before BCT), with external radiotherapy in 8 (1 before BCT) and with both in 2. The radioactive sources were ^{192}Ir in 26 patients and ^{125}I in 4. The sources were implanted permanently in 28 patients and temporary in 2. The tumour peripheral dose given by BCT in permanent implants ranged between 70–100 Gy in 19 cases and between 100–130 Gy in 9; as for temporary implants the given peripheral dose was 40–60 Gy in 5–7 days (8–12 Gy/24 h).

Results

Slow Growing Tumours

The follow-up ranged between 0.3 and 6.9 years (m = 2.5 years). The duration of symptoms before treatment ranged between 0.5 and 9.2 years (m = 3.9 years).

Survival. The five-year survival rate for the 25 patients with slow growing tumours was 88% (± 6.5%) (Fig. 1). Two patients of this series died within the first

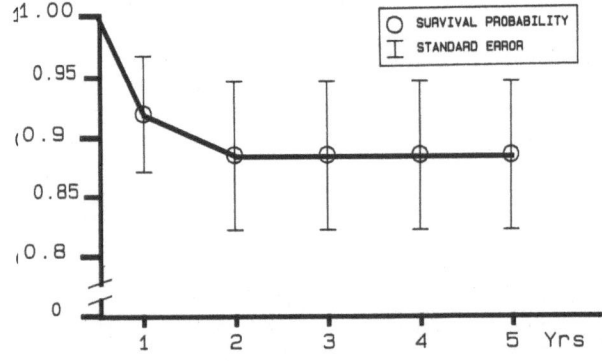

Fig. 1. Survival rate of slow growing tumours at 5 years (88% ± 6.5%) (n = 25)

year and 1 within the second year since BCT: however the death of these 3 patients was not related to the treated tumour (brain infarction contralateral to the implanted side, acute thyreotoxic syndrome and suicide respectively). All the other patients of this series are still alive, but 1 who died owing to tumour progression 5.7 years after interstitial irradiation.

Performance status. Analysis of the Karnofsky performance status since the beginning of BCT and independent of any preceding treatment, was limited to the 11 cases with at least 3 years follow-up. The initial score ranging between 0.60 and 0.70 increased to 0.80–1.00 in all patients during the first year after treatment. This improvement was lasting for many of them, a subsequent decrease of the scale rate being evident in 4 patients who developed radionecrosis. The final score at 3 and 5 years follow-up however never fell below 0.60, ranging in most of the treated cases between 0.70 and 0.90.

Effect on epilepsy. Epilepsy is the most common symptom in slow growing tumours[5, 10]. All patients but one included in our series complained of seizures $^{24}/_{25}$. Their frequency in patients surviving 3 and 5 years decreased after treatment and to zero in 2 patients. In most cases the effect of interstitial irradiation on seizures was early evident and remained stable in the time.

Malignant Tumours

General follow-up ranged between 8 and 35 months (m = 33 months). Clinical history before treatment lasted between 3 and 15 months (m = 7 months).

Survival. The 3 anaplastic tumours grade III (2 astrocytomas and 1 oligodendroglioma) were all living at 2 years follow-up (mean survival = 31 months). Glioblastomas lived respectively 8 and 12 months after interstitial irradiation (Table 1).

Performance status. The Karnofsky scale showed an improvement of the quality of life in the 4 patients evaluated 1 year after treatment, and no decrease of the performance status after 3 years.

Table 1. *Survival of Malignant Tumours (n = 5)*

Grade III	27 months
(n = 3)	32 months
	35 months
Grade IV	8 months
(n = 2)	12 months

Tolerance of BCT

There was no mortality nor morbidity related to the stereotactic implant. Radionecrosis occurred in 9 patients, with a latency of 9–54 months (m = 20 months). Corticosteroids were sufficient to control it in 5 cases. In the remaining 4 patients surgery was indicated because of the extent of radionecrosis (2 cases) and because of tumour progression (2 cases) as documented with stereotactic biopsy.

Discussion

Although our experience is too short and too small to allow definite conclusions, the following comments can be made.

Slow growing tumours require long-term follow-up for the assessment of results. The majority of series reporting the outcome of such types of tumour after surgery plus conventional radiotherapy indicate a survival rate not superior to 30–40% of treated patients at 5 years[6].

For malignant tumours a mean survival rate for anaplastic astrocytoma of 31 months seems noteworthy to us. It is noteworthy that BCT was the sole specific therapy in all cases but one who received in addition an external radiotherapy.

Tumour necrosis should represent the final goal of BCT, giving the best chance of cure[1, 3, 7, 10]. Obviously it has to be carefully balanced with brain tolerance and with the risk of healthy brain injury. Though often unpredictable, it seems to some extent related to the dose/volume and dose/time ratio and to the biopathological characteristics of the tumour.

The quality of life was improved in all patients of our series and remained satisfactory for long time; in malignant tumours the final period of deterioration was short.

To conclude, our data are consistent with those of wider series[2–4, 10] in indicating that stereotactic BCT can play a useful role in the treatment of brain tumours and represents a field in continuous evolution. The choice of the sources, their modalities of application, the dose/time and dose/volume ratio, the place of the treatment in comparison or in association or in association with other therapeutic means are the major problems to be faced in the coming future.

Acknowledgements

This research is partially supported by a grant of Ministry of Public Education of Italy.

References

1. Bernstein M, Gutin PhH (1981) Interstitial irradiation of brain tumours: a review. Neurosurgery 9: 741–750
2. Gutin PhH, Bernstein M (1984) Stereotactic interstitial brachytherapy for malignant brain tumours. Prog Exp Tumour Res 28: 166–182
3. Mundinger F (1987) Stereotactic biopsy and technique of implantation (instillation) of radionuclids. In: Jellinger K (ed) Therapy of malignant brain tumours. Springer, Wien New York, pp 134–194
4. Mundinger F, Weigel K (1984) Long-term results of stereotactic interstitial curietherapy. Acta Neurochir (Wien) [Suppl] 33: 367–371
5. Rossi GF, Scerrati M, Roselli R (1985) Epileptogenic cerebral low-grade tumours: effect of interstitial stereotactic irradiation on seizures. Appl Neurophysiol 48: 127–132
6. Sauer R (1987) Radiation therapy of brain tumours. In: Jellinger K (ed) Therapy of malignant brain tumours. Springer, Wien New York, pp 195–276
7. Scerrati M, Arcovito G, D'Abramo G, Montemaggi P, Pastore G, Piermattei A, Romanini A, Rossi GF (1982) Stereotactic interstitial irradiation of brain tumours: preliminary report. RAYS (Roma) 7: 93–99
8. Scerrati M, Rossi GF (1984) The reliability of stereotactic biopsy. Acta Neurochir (Wien) [Suppl] 33: 201–205
9. Scerrati M, Rossi GF, Roselli R (1987) The spatial and morphological assessment of cerebral neuroectodermal tumours through stereotactic biopsy. Acta Neurochir (Wien) [Suppl] 39: 28–33
10. Szikla G, Schlienger M, Blond S, Daumas-Duport C, Missir O, Miyahara S, Musolino A, Schaub C (1984) Interstitial and combined interstitial and external irradiation of supratentorial gliomas. Results in 61 cases treated 1973–1981. Acta Neurochir (Wien) [Suppl] 33: 355–362

Correspondence: Massimo Scerrati, M.D., Istituto di Neurochirurgia, Università Cattolica S. Cuore, Largo A. Gemelli, 8, I-00168 Roma, Italy.

Vascular Diseases

Acta Neurochirurgica, Suppl. 46, 99–101 (1989)
© by Springer-Verlag 1989

Is Vasodilatation Following Dorsal Column Stimulation Mediated by Antidromic Activation of Small Diameter Afferents?*

B. Linderoth[1], **I. Fedorcsak**[2], and **B. A. Meyerson**[1]

[1] Department of Neurosurgery, Karolinska Hospital, Stockholm, Sweden and [2] National Institute of Neurosurgery, Budapest, Hungary

Summary

It is well-known that high intensity electrical stimulation of peripheral nerves and dorsal root fibres causes vasodilatation. However, the mechanisms underlying the vasodilatory effect of stimulation applied to the dorsal columns (DCS), with an intensity insufficient to recruit small diameter, high threshold fibres are virtually unknown. The present project was planned to elucidate underlying neural mechanisms.

Albino rats, anaesthetized, paralyzed and artificially ventilated were used. Electrical stimulation with different parameters was applied to various sites of the exposed spinal cord, root fibres and peripheral nerves. In some experiments the spinal cord, root fibres or peripheral nerves were transected. Peripheral blood flow was recorded using laser Doppler technique.

With stimulation of the the lower thoracic region at low intensity substantially increased blood flow in the ipsilateral hind paw. The compound action potentials from the gural nerve displayed only components from low threshold, rapidly conducting fibres without detectable late components. Transection of the spinal cord above the stimulation site did not affect the blood flow changes. Also low intensity stimulation of the proximal part of a sectioned dorsal root resulted in a substantial rise in peripheral blood flow, whereas the same intensity proved ineffective when applied to the distal stump. High intensity stimulation of the distal stump not unexpectedly, caused a major increase in blood flow.

The findings in the present study implicate the importance of a central circuit for the effect, whereas antidromic activation of primary afferents seems to be a less likely explanation. Possibly, stimulation induces a transitory inhibition of sympathetic vasoconstrictor tone, though activation of sympathetic vasodilatory efferents cannot be excluded.

Keywords: Dorsal column stimulation; vasodilatation; antidromic activation; rat model.

Introduction

Following the publication of a paper by Cook *et al.* in 1975[5] many reports have appeared about the use-fulness of spinal cord stimulation in peripheral vascular disease. It augments peripheral blood flow and skin temperature, promotes ischaemic ulcer healing and even seems to have a limb saving effect[1, 4, 6, 8, 11]. In general, the clinical results have been favourable and ischaemic pain due to peripheral vascular disease has become a major indication for this treatment modality.

The mechanisms underlying the effect of spinal cord stimulation on ischaemic pain and peripheral circulation as well as on neurogenic pain remain obscure. However, several possible mechanisms have been discussed: antidromic stimulation of primary afferent fibers[5, 8], alteration of autonomic nervous system function *e.g.* the inhibition of sympathetic activity (*e.g.*[1, 13]), release of vasoactive substances (*e.g.*[1]), and finally that the beneficial effect on peripheral blood flow might be secondary to the reduction of ischaemic pain itself[1, 4].

The idea of antidromic activation of slow afferent fibers as a possible explanation, as discussed by *e.g.* Groth 1985[6] and by several authors at an international symposium in 1987, is based on an observation already made at the turn of the century by Bayliss[2]. He demonstrated local peripheral vasodilatation in response to high intensity dorsal root stimulation activating thin fibers. This finding was confirmed in 1980 by Hilton and Marshall[7] who found that a rise in muscle blood flow by 50–60% could be provoked by high intensity stimulation of dorsal roots.

The present project was planned to elucidate, in an experimental model, the mechanisms behind the vasodilator effects of dorsal column stimulation. As a first step we investigated the possibility that the antidromic activation of the afferent systems described above, might be a component in the effect on peripheral flow of the stimulation as it is used on patients.

* Supported by grants from the Wennergren Foundation, the Karolinska Institute and Medtronic Inc.

DCS - blood flow in hind paw

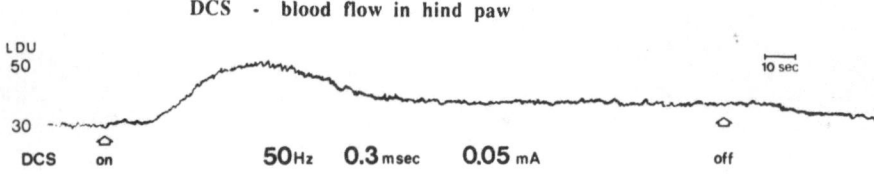

Fig. 1. Skin blood flow in a hind paw during stimulation of the ipsilateral dorsal column. In the experiment illustrated, constant current technique was used. Stimulus on and off is indicated by arrows. Blood flow is expressed in LDU (arbitrary units of the laser Doppler meter)

Material and Methods

Albino rats were anaesthetized with intraperitoneal pentobarbital (6 mg/100 g). The animals were tracheotomized, paralyzed by i.v. gallamine (0,4 mg/100 g) and artificially ventilated monitoring end-tidal pCO_2. Body temperature was maintained at normal level and systemic blood pressure monitored. The dorsal aspect of the spinal cord was exposed by a laminectomy from T 10 to S 1. The dura was opened and the medulla and roots covered with paraffine oil. Stimulation frequency was varied between 0.5–50 Hz; pulse width between 0.05–0.3 msec. Constant voltage technique was generally used, the intensity varied between 0.05–1.0 V. Compound action potentials were recorded peripherally from the sciatic or sural nerves in order to monitor the type of peripheral fibres antidromically activated. A Medelec MS 92 A (Vickers Health Care Company, England) was used for averaging the responses. Peripheral skin blood flow was recorded from a hind paw using laser Doppler technique (Periflux PF 1; Perimed KB, Stockholm, Sweden).

Results

The threshold for a local motor response was found to be approx. 0.7 V (cf.[3]). Fig. 1 illustrates the effect of intradural monopolar stimulation unilaterally of the dorsal column at the level of T 13-L 1. By use of stimulation parameters similar to those used clinically, it was often possible to produce an increase in peripheral flow which initially amounted to 65% or more. After some time (in this case after 50 seconds) the flow decreased to a plateau at least 30% above base line.

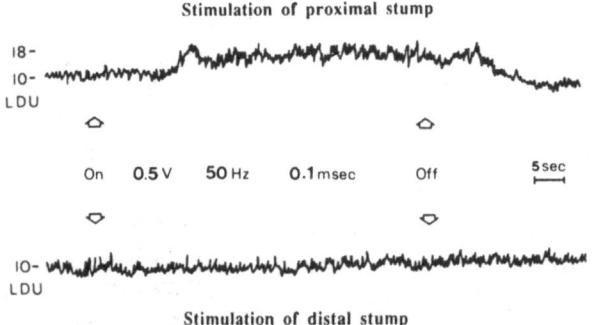

Fig. 2. Stimulation of sectioned dorsal roots innervating the hind paw. Blood flow is expressed in arbitrary units. In this experiment constant voltage was used

Following stimulus off there is a slow decrease to the prestimulation value. However, in several animals, we were unable to demonstrate this effect although all monitored vital parameters were normal. Nevertheless, in all animals it was possible to provoke vasodilatation by using high intensity stimulation.

It could also be demonstrated that local vasodilatation in the hind limb following low intensity stimulation was not related to a general increase in blood pressure.

After cutting the dorsal root fibres to the hind limb, a local peripheral vasodilatation could still be evoked by low intensity stimulation of the ipsilateral dorsal column. For example stimulation using 0.1 V, 50 Hz and 0.1 msec resulted in a flow increase of more than 100% above resting value.

In Fig. 2 the effect of low intensity dorsal root stimulation is demonstrated. The dorsal roots to a hind limb were sectioned and stimulation applied to the proximal as well as to the distal stumps with the same parameters resulting in vasodilatation when applied directly to the dorsal column.

It is evident that only stimulation of the proximal stump provoked an increase in peripheral flow. To be able to produce a vasodilatation by stimulating the distal stump, the stimulus intensity had to be increased to levels where high threshold thin fiber systems could also be recruited.

During stimulation of dorsal columns and roots, the averaged compound action potentials in the sciatic or in the sural nerves were monitored. Only activation of low threshold fast fibres, corresponding to the initial deflections without detectable late components, were recorded when using stimulation intensities sufficient to produce peripheral vasodilatation.

Sectioning of the ipsilateral sciatic nerve, not unexpectedly, caused a major rise in peripheral flow and subsequently no further additions could be recorded with spinal cord stimulation. Furthermore, transection of the spinal cord approx. 15 mm rostrally to the stimulation site did not affect the rise in peripheral blood flow following stimulation.

Discussion

From our experiments we can conclude that the albino rat may serve as a model for the study of peripheral vascular effects of low intensity dorsal column stimulation. Furthermore, it is evident that the vasodilatation can be demonstrated without the activation of high threshold fibre systems (cf.[7, 10]). The increase in peripheral blood flow following stimulation remains after dorsal root transection, a finding incompatible with the idea that antidromic stimulation of primary afferents would play a major role for the effect. Instead, the effect of low intensity stimulation appears to depend on a central, spinal cord mechanism since only centripetal and not centrifugal low intensity stimulation proved effective.

The pain suppressing effect of dorsal column stimulation has been proposed to involve supraspinal loops (*e.g.*[9, 12]). However, in the present experiments the stimulation-induced peripheral vasodilatation could also be demonstrated when the connections to supraspinal centers were eliminated. Besides, vasodilatation occurred only on the side ipsilateral to the dorsal column stimulated and this makes the activation of local spinal reflexes probable.

The effector mechanisms in low intensity dorsal column stimulation are probably quite different from those operating in high intensity dorsal root stimulation. It is probable that modulation of sympathetic activity is of crucial importance. The putative autonomic mediation of vasodilatation following dorsal column stimulation will be discussed in another paper in preparation.

References

1. Augustinsson LE, Holm J, Carlsson CA, Jivegård L (1985) Epidural electrical stimulation in severe limb ischaemia. Evidences of pain relief, increased blood flow and a possible limb-saving effect. Ann Surg 202: 104–111

2. Bayliss WM (1901) On the origin from the spinal cord of the vasodilator fibres of the hind limb and on the nature of these fibres. J Physiol (London) 26: 173–210

3. Broggi G, Franzini A, Parati E, Parenty M, Flauto C, Servello D (1985) Neurochemical and structural modifications related to pain control induced by spinal cord stimulation. In: Lazorthes Y, Upton ARM (eds) Neurostimulation: An overview. Futura Publ Comp, New York, pp 87–95

4. Broseta J, Garcia-March G, Sanchez MJ, Goncales J (1985) Influence of spinal cord stimulation on peripheral blood flow. Appl Neurophysiol 48: 367–370

5. Cook AW, Oygar A, Baggenstos P, Pacheco S. Kleriga E (1976) Vascular disease of extremities. New York State J Med 76: 366–368

6. Groth KE (1985) Spinal cord stimulation for the treatment of peripheral vascular disease. In: Fields HL *et al* (eds) Advances in pain research and therapy, vol 9. Raven Press, New York, pp 861–870

7. Hilton SM, Marshall JM (1980) Dorsal root vasodilatation in cat skeletal muscle. J Physiol (London) 299: 277–288

8. Illis LS, Sedgwick EM, Tallis RC (1980) Spinal cord stimulation in multiple sclerosis: clinical results. J Neurol Neurosurg Psychiatry 43: 1–14

9. Larson SJ, Sances A, Riegel DH, Meyer GA, Dallmann DE, Swiontek T (1974) Neurophysiological effects of dorsal column stimulation in man and monkey. J Neurosurg 41: 217–223

10. Magerl W, Szolcsányi J, Westerman RA, Handwerker HO (1987) Laser Doppler measurements of skin vasodilation elicited by percutaneous electrical stimulation of nociceptors in humans. Neurosci Lett 82: 349–354

11. Meglio M, Cioni B, Dal Lago A, De Sandis M, Pola P, Serricchio M (1981) Pain control and improvement of peripheral blood flow following epidural spinal cord stimulation. J Neurosurg 54: 821–823

12. Saadé NE, Tabet MS, Soueidan SA, Bitar M, Atweh SF, Jabbur SJ (1986) Supraspinal modulation of nociception in awake rats by stimulation of the dorsal column nuclei. Brain Res 369: 307–310

13. Tallis RC, Illis LS, Sedgwick EM, Hardwidge C, Garfield JS (1983) Spinal cord stimulation in peripheral vascular disease. J Neurol Neurosurg Psychiatry 46: 478–484

Correspondence: Dr. B. Linderoth, Department of Neurosurgery, Karolinska Hospital, S-104 01 Stockholm, Sweden.

Acta Neurochirurgica, Suppl. 46, 102–104 (1989)
© by Springer-Verlag 1989

Cerebral and Carotid Haemodynamic Changes Following Cervical Spinal Cord Stimulation. An Experimental Study

G. García-March, M. J. Sánchez-Ledesma, J. Anaya, and **J. Broseta***

Department of Neurosurgery, Hospital Virgen de la Vega, Universidad de Salamanca, Salamanca, Spain

Summary

Since it is accepted that spinal cord stimulation may produce segmentary vasodilation, it is presumable that when applied in the high cervical segments some carotid and cerebral blood flow changes can be expected. Following this assumption, 25 dogs and 25 goats were used. Under routine experimental conditions a C 7 laminectomy was performed in these animals and a bipolar lead introduced and manipulated in the epidural space till the right C 2 segment. Right common and internal carotid arteries of the dogs were isolated and electromagnetics probes placed for continuous monitoring of blood flow changes. Right internal maxillary artery was isolated and its branches ligated for flowmetry of hemispheric blood flow in the goat. [131]I antipyrine also studied to control regional cerebral blood flow changes. Arterial pressure and blood gasometry were periodically determined to avoid masking results. Pulse width of 0.1 to 0.2 msec, 80 to 120 cps and amplitude to muscle contraction threshold at low rate were used as electrical parameters. After stimulation common and internal carotid blood flow increased with a mean of 60% and hemispheric blood flow with a mean of 55% according to flowmetry findings. Iodoantipyrine studies showed an average increase of 35%. These changes were not modified by atropine, morphine and naloxone and partially blocked by indomethacin, cimetidine and propanolol.

Keywords: Spinal cord stimulation; carotid blood flow; cerebral blood flow; goat.

Introduction

Today it is accepted that spinal cord stimulation may induce a segmental vasodilator response. Clinical reports show haemodynamic evidence that stimulation increases peripheral[4–7, 15, 17] and coronary[11, 14] blood flow when applied in the lower or higher spinal cord segments respectively. The mechanisms by which stimulation at these levels increases blood flow are unknown though a vasodilation secondary to a sympatholytic

effect[3, 6, 9] or the mediation of local vasoactive substances[6, 8, 17, 18] have been basically postulated.

With this and other experimental data[1, 2, 13] we considered the possible haemodynamic effect of this technique when electrical impulses are given in the region of the spinal autonomic segmentary control for carotid and cerebral circulation. An experimental protocol was devised in which electrical stimulation was applied in the right C_1–C_2 cord segments in 25 dogs and 20 goats. Blood flow was measured in the right carotid arteries and in both cerebral hemispheres using electromagnetic flowmetry and ^{131}I-antipyrine cerebral scintigraphy.

Experimental Material and Method

Twenty-five adult mongrel dogs weighing 15 to 28 kg were used to determine with electromagnetic flowmetry the haemodynamic changes in the carotid territory under stimulation conditions. These animals were not used for cerebral blood flow studies because of the small contribution of the internal carotid artery to brain circulation. Therefore, 20 adult goats weighing 25 to 45 kg were used to investigate this aspect using flowmetry of the internal maxillary artery, which supplies most of the hemisphere blood flow when its collateral branches are ligated via a *rete mirabile*[16], and iodoantipyrine brain scintigraphy respectively.

For surgical manœuvres, general anesthesia was induced by intramuscular injection of pentobarbital (40 mg/kg), and maintained with endotracheal intubation, mechanical ventilation and pentobarbital (30 mg/kg) and atropine (0.05 mg/kg) i.v. infusion. The right femoral artery was cannulated for continuous arterial blood pressure monitoring. Blood gasometry was determined every 30 minutes. For spinal cord stimulation, in the dog as well as in the goat, a bipolar lead was introduced in the epidural space through a small laminectomy and manipulated under fluoroscopy to the highest cervical cord segments homolateral to the explored vascular territory, since the most evident haemodynamic changes on common carotid and internal maxillary arteries following stimulation were found at that side and level as shown in a previous study (Fig. 1). With the electrode fixed, the right common and internal carotid arteries of the dog were

---

* J. Broseta, M.D., Cátedra de Neurocirugía, Facultad de Medicina, c/o Espejo, E-37007 Salamanca, Spain.

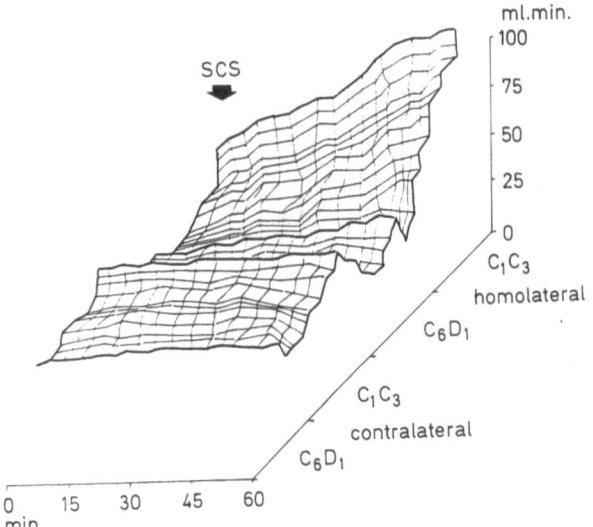

Fig. 1. Right common carotid blood flow changes during 60 minutes of cervical spinal cord stimulation with positions of the electrode

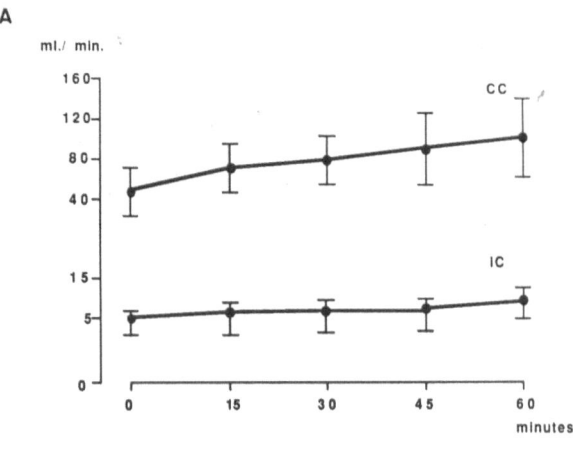

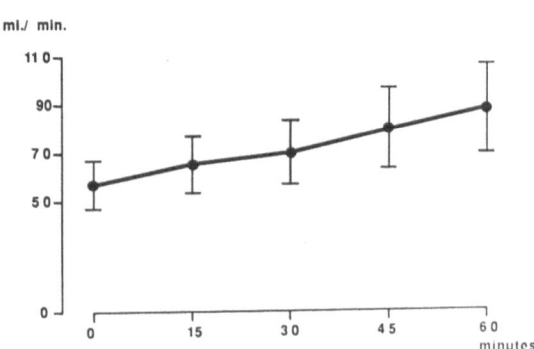

Fig. 2. Mean changes in blood flow during a period of 60 minutes of C_1–C_2 spinal cord stimulation homolateral to the explored vessel. A) Measures in the right common carotid (*CC*) and internal carotid (*IC*) arteries. B) Recordings in the right internal maxillary artery

isolated and an appropriate electromagnetic probe was placed around them. In the case of the goat, the maxillary artery was isolated after partially removing the jaw and ligation of its branches to avoid extracranial blood flow contamination after which the probe was placed around. After 30 minutes, stimulation was given for 60 minutes on/off/on intervals (pulse duration of 0.1 to 0.2 msec, a rate of 80 cps). The amplitude was kept to subliminal fasciculation threshold of the homolateral cervical paravertebral and facial musculature. Flowmetry was continuously recorded throughout the experience.

^{131}I-antipyrine brain scintigraphy was also used in the goat for qualitative cerebral blood flow measurements. Once the lead was placed and fixed in the proper position and the cephalic vein isolated, pericraneal structures and temporal muscles were removed to reduce extracranial activity. The animal was moved to the gammacamara for exploration. One ml of antipyrine with 1 mCi of ^{131}I was intravenously injected before and 60 minutes following stimulation. The activity time curves were analyzed by computer that recorded the hemisphere and regional arterial and capillar distribution as well as the arteriocapillary angle to compare changes in blood flow.

Results

Electromagnetic recordings from common and internal carotid arteries showed an average increase in blood flow of 66 and 63% respectively after 60 minutes of high cervical spinal cord stimulation. On basal levels of 59.6 ± 23.6 ml/min in the common carotid artery and of 5.1 ± 2.3 ml/min in the internal carotid one, post-stimulation measures of 99.1 ± 39.6 ml/min ($p < 0.01$) and 8.3 ± 3.7 ml/min ($p < 0.01$) were obtained respectively (Fig. 2 A). The highest increase in blood flow appeared during the first 15 minutes of stimulation. In the group with 1 hour on/off/on periods,

the increased blood flow fell during the off phase although still raised.

Flowmetry of the internal maxillary artery showed an average increased in hemisphere blood flow of 55% after 60 minutes of stimulation with values of 89 ± 17.5 ml/min from previous 57.2 ± 9.8 ml/min ($p < 0.01$) (Fig. 2 B). In this case the highest increase in blood flow occurred during the first 15 minutes of stimulation and between 30 and 45 minutes.

Iodoantipyrine brain scintigraphy analysis showed a significant increase of activity in the hemisphere homolateral to the stimulation side. This was manifested by a higher activity during the arterial phase and a flater capillary line indicating that radioactive material remained longer in circulation expressed by a opening of the arteriocapillary angle (Fig. 3). Increase in regional blood flow were also detected in the homolateral frontal area, partially in both parietal area and was absent in posterior fossa.

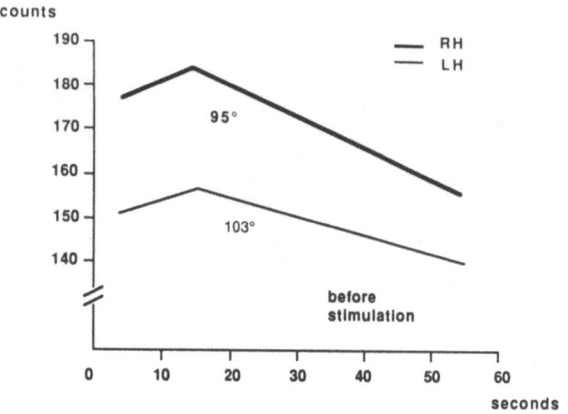

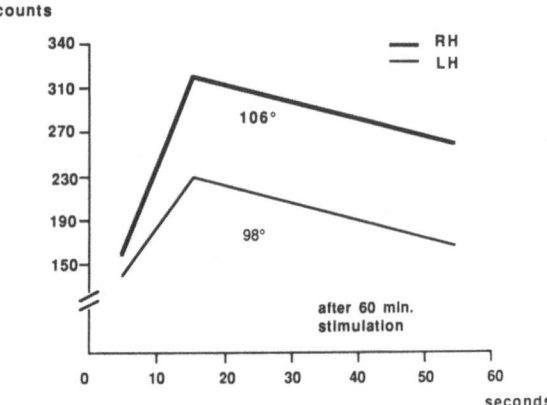

Fig. 3. Changes in ^{131}I-antipyrine brain scintigraphy before and 60 minutes following stimulation recorded at the right (*RH*) and left (*LH*) hemispheres. See text

Discussion

This study suggests that high cervical spinal cord stimulation with these quoted electrical parameters induced significant cerebral and carotid blood flow increase. These findings are qualitative and quantitative similar to those described by other groups. Using the ^{133}Xe clearance Hosobuchi[10] found a 34% increase of blood flow in the hemisphere homolateral to the stimulation side in patients. Recently, Kanno *et al.*[12] observed a global cerebral blood flow increase of 20 to 40% in patients in a vegetative state treated with C_2 spinal cord stimulation to alleviate spasticity.

From these observations and since spinal cord stimulation is harmless, it seems rational to suggest that it be used for such cerebrovascular disorders as vasospasm, diffuse cerebral atheromatosis or cerebral ischaemia unresponsive to conservative treatment.

Acknowledgments

We are fully indebted to Concha Blasco and Fran Catalán for their valuable assistance and kindly attentions during the development of this study.

References

1. Alborch E, Gómez B, Dieguez G *et al* (1977) Cerebral blood flow and vascular reactivity after removal of the superior cervical sympathetic ganglion in the goat. Circ Res 41: 278–282
2. Aubineau PF, Sercombe R, Seylaz J (1975) Continuous recordings of local cerebral blood flow during cervical sympathetic nerve blockade or stimulation. J Physiol 246: 104–106
3. Augustinsson LE (1981) Discussion on spinal cord stimulation and peripheral blood flow. In: Hosobuchi Y, Corbin T (eds) Indications for spinal cord stimulation. Excerpta Medica, Amsterdam, pp 72–75
4. Augustinsson LE, Carlsson CA, Fall M (1982) Autonomic effects of electrostimulation. Appl Neurophysiol 45: 185–189
5. Broseta J, Garcia-March G, Sánchez-Ledesma MJ *et al* (1985) Influence of spinal cord stimulation on peripheral blood flow. Appl Neurophysiol 48: 367–370
6. Broseta J, Barberá J, De Vera JA *et al* (1986) Spinal cord stimulation in peripheral arterial disease. A cooperative study. J Neurosurg 64: 71–80
7. Fiume D (1983) Spinal cord stimulation in peripheral vascular pain. Appl Neurophysiol 46: 290–294
8. Grönstad KO, Ahlman H, Zinner MJ *et al* (1983) The effect of vagal nerve stimulation of feline portal venous levels. In: Skrabanek P, Powell D (eds) Substance P. Dublin, pp 153–154
9. Hilton SM, Marshall JM (1980) Dorsal root vasodilatation in cat skeletal muscle. J Physiol (Lond) 299: 277–288
10. Hosobuchi Y (1985) Electrical stimulation of the cervical spinal cord increases cerebral blood flow in humans. Appl Neurophysiol 48: 372–376
11. Illis LS, Sedgwick EM, Tallis RC (1980) Spinal cord stimulation in multiple sclerosis: clinical results. J Neurol Neurosurg Psychiatry 43: 1–14
12. Kanno T, Kamei Y, Yokoyama T (1987) Effects of neurostimulation on neuronal reversibility-experience of treatment for vegetative status. Presented to the 8th Congress of the European Association of Neurological Societies. Barcelona
13. Lluch S, Vallejo AR, Dieguez G *et al* (1978) Adrenergic involvement in cerebral blood flow changes in controlled hypotension. In: Cervos-Navarro J (ed) Advances in neurology, vol XX. Raven Press, New York, pp 215–221
14. Mannheimer C, Carlsson CA, Wilhelmsson C (1982) Transcutaneous electrical nerve stimulation in severe angina pectoris. Europ Heart J 3: 297–302
15. Meglio M, Cioni B (1982) Personal experience with spinal cord stimulation in chronic pain management. Appl Neurophysiol 45: 195–200
16. Reimann C, Lluch S, Glick G (1972) Development and evaluation of an experimental model for the study of the cerebral circulation in the unanaesthetized goat. Stroke 3: 322–328
17. Tallis RC, Illis LS, Sedgwick EM *et al* (1983) Spinal cord stimulation in peripheral vascular disease. J Neurol Neurosurg Psychiatry 46: 478–484
18. Wei EP, Kontos HA, Said Sl (1980) Mechanism of action of vasoactive intestinal polypeptide on cerebral arterioles. Am J Physiol 239: 765–768

Correspondence: J. Broseta, M.D., Cátedra de Neurocirugía, Facultad de Medicina, c/Espejo, E-37007 Salamanca, Spain.

Technical Progress

Acta Neurochirurgica, Suppl. 46, 107–108 (1989)
© by Springer-Verlag 1989

CT-Guided "Real Time" Stereotaxy

H. F. Reinhardt and **H. Landolt**

Neurosurgical University Clinic, Kantonsspital, Basel, Switzerland

Summary

Starting from a previously developed 3-D digitizer for a CT correlated topographic orientation during open brain surgery a computerized targeting device was constructed specially designed for stereotaxy. CT and later MRI data are transferred to an IBM compatible personal computer. The spatial information is registered by a multi-axis measuring arm. Advantages and future development of computerized stereotaxy are discussed.

Keywords: Computerized stereotaxy; CT-MRI-guided stereotaxy; targeting device; 3-D digitizer; brain neoplasm.

Introduction

Between 1983 and 1986 at the University of Basel, Switzerland a computerized measuring device with a 4-axis digitizing arm[3] was developed for open brain tumour operations. With its aid it was possible to correlate the actual intracranial location with preoperative CT scans. At the same time as a Japanese team[4], we independently became aware of the possibility of aiming at a target in the absence of direct contact with structures in the measuring field. The computer is not able to distinguish whether the solid end-segment of the measuring arm is present or not. Based on this principle we were able to realize a "real time"-stereotaxy surprisingly simple to manipulate-provided however, that precision could match the standards of conventional stereotactic frames.

Material and Methods

Based on a 3-D graphic digitizer for the Apple-2 e PC[1] and the aforementioned experimental device for use in open surgery[2, 3] we recently constructed a high precision measuring device (Fig. 1) exclusively designed for stereotaxy. The arm has 4 articulations and one extensible axis; moreover, it can be fixed in 5 positions of the halo-ring of the headholder. In this way $5^1/_2$ degrees of freedom are achieved. Because of the relative smallness and quasi-infinite resolution precision potentiometers were preferred to optical encoders or inductive resolvers. The whole folded arm weighing 3,900 g can be stored in a container of $36 \times 18 \times 12$ cm.

For position calculation and imaging procedures a fast 286 IBM-AT clone with mathematical co-processor is used (Fig. 2). The 5 resistance values are digitized in a 12-bit AD converter, the spatial information is computed in polar coordinates and then transformed into Cartesian coordinates; this extensive trigonometry is done in 40 msec. For lack of better interfaces the image data from the Siemens-DRH CT scanner must be transferred to the hard disc of the IBM compatible computer by 8″ floppy discs.

X- and y-axes are represented by a cursor-cross, the z-axis corresponds to the axial plane of the CT pictures. The 5 to 40 pictures are turned over like a staple of photographs at the position of the measuring tip in the z-axis, so that adequate picture of the target zone is always visible. The pictures are displayed in selectable false-colours in the original 256×256 matrix.

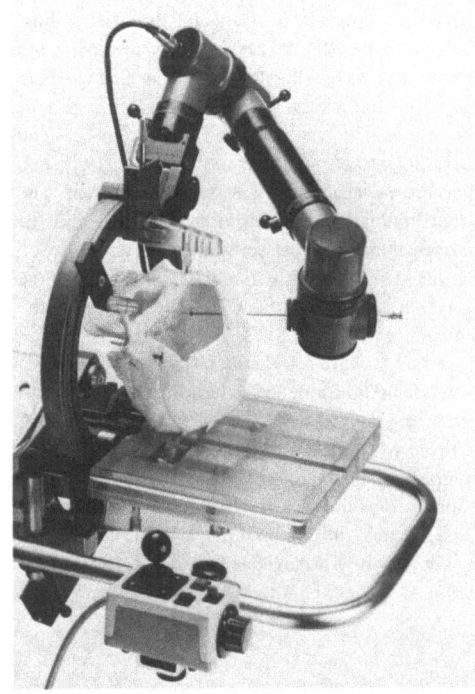

Fig. 1. Digitizing arm on halo-ring headholder. Bottom: control-panel

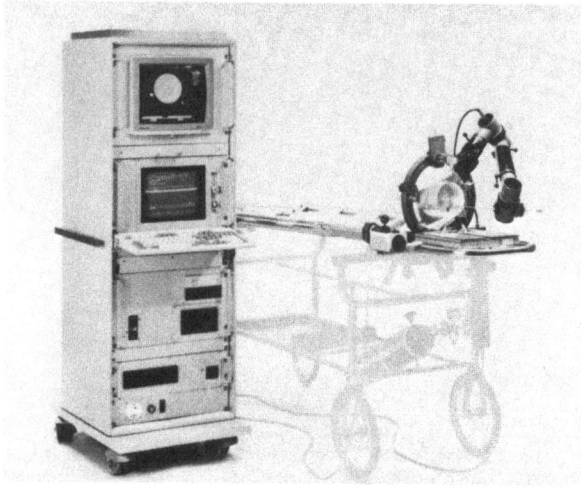

Fig. 2. IBM-AT compatible computer in 19″ rack (left side). Showel stretcher-headholder unit with measuring arm (right side)

The patient lies on a vacuum mattress fixed to a showel stretcher on which a heavy halo-ring is rigidly attached (Fig. 2, right side). The head is fixed with radiolucent polycarbon-pins. The whole system is connected to the CT table and precisely centered.

Starting from a lead-mark in the base plane close to the haloring CT images are recorded, generally in 4 mm slices. The data are then transferred to the IBM compatible PC, where they are reformatted for use in PC-DOS. During this procedure the patient is transported in an *unchanged position* from the CT suite to the operating theatre.

The measuring arm is fixed at the optimum position onto the haloring; after a short position check and eventual calibration the arm and the small control-panel (Fig. 1, bottom) are wrapped in sterile plastic bags or sheets. Without any needle or tool and using only the virtual axis of the end-segment the best site for the burr hole and target can be determined by freely moving the articulations, which are finally blocked by efficient brakes. After loosening the brakes of one axis the arm is temporarily turned aside and the burr hole is made with entirely free access. Repositioning is done without new targeting procedures with the aid of an acoustic signal. The rigid arm now serves as a stable stand for all sorts of instruments, *e.g.* biopsy-needle, forceps, endoscope, measuring probes etc. The length of the different graduated instruments is fed into the computer by the small control-panel where the software menu can also be manipulated by a joy-stick. For the pathologist who inspects the frozen sections, a colour hardcopy of the CT picture indicating the site of biopsy is made.

For security reasons we have attached a base-plate of the conventional target centered Patil-frame to the head-holder in order to be able to continue stereotaxy even in case of technical failure of the new system. From May 1988 our device has tested clinically. Accuracy lies within ± 1–2 mm verified by repetitive CT scan controls. Principal applications are tumour biopsies and endoscopic evacuations of deep-seated intracerebral haematomas. A head-holder-/stretcher-unit in non-ferromagnetic materials for MRI is at present being built.

Discussion

In 1983 – the onset of our development in computerized CT-guided targeting devices[2, 3] – the implica-

tions of computer-aided CT-guided stereotaxy could hardly be imagined. At present the cost for hi-tech is perhaps not justified by the clinical usefulness, but the advantages of computerized stereotaxy are nevertheless becoming evident:

– the actual tip position of the probe can be permanently observed in real-time;

– burr hole and target can freely be chosen without time-consuming calculations;

– different targets can be reached through the same burr hole;

– additional software routines can be integrated directly into the stereotactic program:

– X-ray therapy planning: isodoses, seed coordinates; spatial data: MRI, digital subtraction angiography (DSA); EEG-mapping etc; brain atlas or landmarks; 3 D-imaging; robotics.

Computer-assisted stereotaxy will become particularly useful when integrated into existing (imaging procedures, radiosurgery etc.) and future methods or devices (handling/roboting tools) based on microprocessors.

In order to get more freedom of movement in open brain surgery and to be able to comply to the very demanding precision of the measuring arm we are currently trying to develop a new series of stereotactic devices which will work with a minimum of mechanical parts. The principal item will be a Sonar digitizer working with miniature positioning probes which emit sound close to the hearing threshold. There are some physical problems like compensation of temperature, air-flow and surrounding noise which probably can be resolved by appropriate sensors and software. Our plan is to develop a small "addendum" to different sorts of tools like needle-carriers for stereotaxy, or handpieces (pointers, sensors, ultrasonic-aspirator/laser etc.) for microsurgery.

References

1. Krupp L (1983) The space tablet (TM), Micro Control Systems Inc., 143 Tunnel Rd. Vernon, Conn. 06066 USA
2. Reinhardt H, Stula D, Gratzl O (1985) Topographic studies with 32-P tumour marker during operations of brain tumours. Eur Surg Res 17: 333–340
3. Reinhardt H, Meyer H, Amrein E (1988) A computer assisted device for the intraoperative CT correlated localization of brain tumours. Eur Surg Res 20: 51–58
4. Watanabe E, Watanabe T, Manaka S *et al* (1987) Three-dimensional digitizer (Neuronavigator): new equipment for CT-guided stereotaxic surgery. Surg Neurol 27: 543–547

Correspondence: PD Dr. Hans F. Reinhardt, Neurochirurgische Universitätsklinik, Kantonsspital, CH-4031 Basel, Switzerland.

Acta Neurochirurgica, Suppl. 46, 109–111 (1989)
© by Springer-Verlag 1989

Transposition of Image-defined Trajectories into Arc-quadrant Centered Stereotactic Systems

L. Zamorano, A. Martinez-Coll, and **M. Dujovny**

Henry Ford Neurosurgical Institute, Department of Neurological Surgery, Henry Ford Hospital, Detroit, MI, U.S.A.

Summary

A methodology and computer program used for transposition of image-defined trajectories into stereotactic space is presented. Arc-quadrant centered stereotactic frames (Leksell, modified by us Komai) are based on the principle of aiming a target point from any entry of a formed sphere of known radius. Frames adapted isocentrically to CT gantry and parallel to the scanning plane allow interactive selection of targets and trajectories based on multiplanar reformatted images. The mathematical fundaments to calculate angle A in the coronal plane and angle B in the saggital plane based on the geometrical configuration of arc-quadrant centered stereotactic devices are presented.

Keywords: Stereotaxis; computed tomography; computer programming; imaging.

Introduction

The aim of any imaging-guided stereotactic system is to transpose image-generated data (coordinates, trajectories, volume) into the stereotactic space. Arc-quadrant centered stereotactic devices *i.e.* Leksell, Hitchcock, Laitinen, etc. (including our modified, Komai) are probably the most commonly used stereotactic devices. The work is based on the principle of aiming a target point from any entry of a formed sphere of known radius. The transposition of trajectories from 3 D/2 D Multiplanar images into stereotactic space is mandatory in many volumetric oriented-stereotactic procedures, like interstitial radiotherapy, endoscopic laser photocoagulation, laser resection, biopsy following axis of tumour, etc. Therefore, we developed a methodology and computer program to calculate angular settings of arc-quadrant stereotactic devices.

Material and Method

Stereotactic Frames and Image-interfacing

Arc-quadrant centred stereotactic devices (Leksell, modified by us Komai) were adapted isocentrically to the CT gantry and parallel

to the plane of scanning (Fig. 1). Isocentricity and parallelism as well as the existence of coronal and sagittal angular scales are mandatory for the methodology and computer program.

Methodology

Once isocentricity and parallelism are confirmed, axial images containing target volume are selected. Target coordiantes are measured directly by using the software capabilities of the GE-9800 CT scanner. X and Y are read directly (grid or region of interest function) and Z is measured directly by the CT table positioning. Reformatted images on coronal, sagittal or oblique paraaxial planes are generated and target points and trajectories are selected and displayed. This methodology allows an interactive selection of targets and trajectories and correction of them based on the multiplanar information. A trajectory is defined by selecting a second point from another axial, reformatted or scoutview images and its coordinates are measured. Final selected targets and trajectories are plotted on multiplanar reformatted images and anteroposterior and lateral scoutviews.

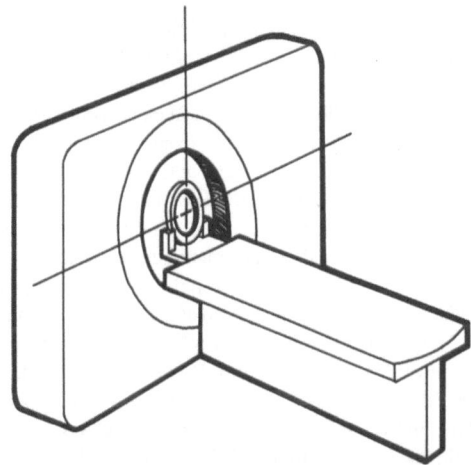

Fig. 1. Stereotactic device isocentric to the CT gantry and in a parallel relationship with the scanning plane

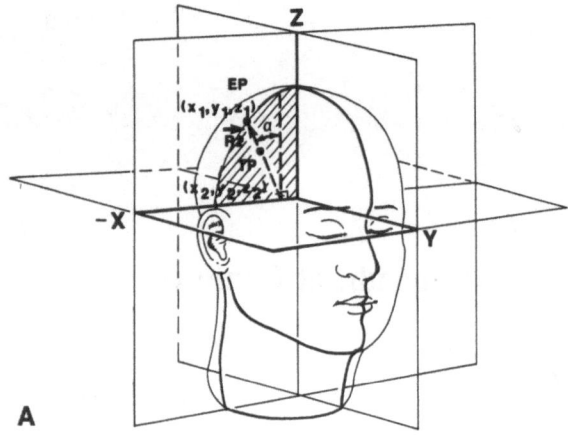

A

Fig. 2. Angle A represents the angle on the X–Z plane

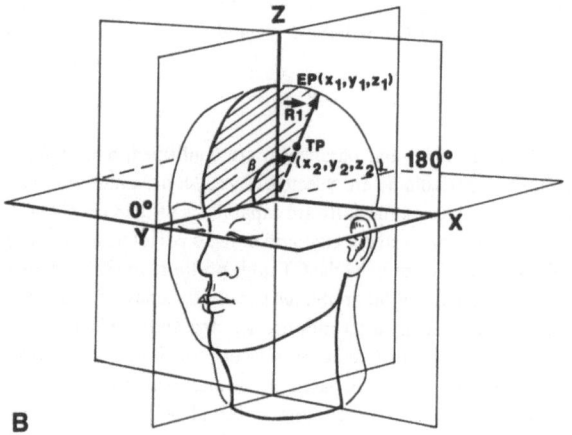

B

Fig. 3. Angle B represents the angle on the Y–Z plane

Computer Program

A computer program to calculate the angles A (X–Z plane) and B (Y–Z planes) in order to transpose preselected CT trajectories into the stereotactic space was written on Basic and Fortran. The input consists of target point coordinates ($x2$, $y2$, $z2$) and coordinates of a second point ($x1$, $y1$, $z1$) calculated on axial, reformatted, or scoutview images to define a trajectory. The output of the program are angles in coronal (A) and sagittal (B) planes (Figs. 2 and 3). The mathematical fundamental of the program are: the distance between two points in space can be found by substracting the coordinates of one point from the coordinates of a second point. This subtraction yields the distance between two points in each of the three dimensions (x, y, z) (eq. 1–3). Once these distances are known (Rx, Ry, Rz) a resultant direction can be calculated (eq. 4–5). Using trigonometric relationships, R 1 is the magnitude of a resultant vector with respect to the Y–Z plane and R 2 is the magnitude of the resultant vector with respect to the XZ plane. Sagittal angle B is defined as the inverse cosine of the cos(B), where cos(B) = Ry/R 1 (eq. 6–8). Angle A is defined as the inverse cosine of cos(A), where cos(A) is the absolute value of Rx/R 2 (eq. 7–9).

The mathematical fundamentals of the program to calculate angles A and B are:

$$Rx = X_1 - X_2 \tag{1}$$

$$Ry = Y_1 - Y_2 \tag{2}$$

$$Rz = Z_1 - Z_2 \tag{3}$$

$$\begin{aligned}R_1 &= \sqrt{Ry^2 + Rz^2} \\ &= \sqrt{(y_1 - y_2)^2 + (Z_1 - Z_2)^2}\end{aligned} \tag{4}$$

$$\begin{aligned}R_2 &= \sqrt{Rx^2 + Rz^2} \\ &= \sqrt{(X_1 - X_2)^2 + (Z_1 - Z_2)^2}\end{aligned} \tag{5}$$

$$\begin{aligned}COS\,B &= \frac{Ry}{R_1} \\ &= \frac{(y_1 - y_2)}{\sqrt{Ry^2 + Rz^2}}\end{aligned} \tag{6}$$

$$\begin{aligned}COS\,A &= \frac{Rx}{R_2} \\ &= \frac{(X1 - X2)}{\sqrt{Ry^2 + Rz^2}}\end{aligned} \tag{7}$$

$$\begin{aligned}B &= arcos\,B \\ &= arcos\frac{Ry}{R_1}\end{aligned} \tag{8}$$

$$\begin{aligned}A &= arcos\,A \\ &= acros\frac{Rx}{R_2}\end{aligned} \tag{9}$$

Accuracy Testing

Phantom measurements of accuracy and neproduubility were verified by a series of tests: 1 mm titanium targets were placed on an adult human skull fixed to the frame in different positions. Trajectories were defined by selecting a second point in the stereotactic space. Coordinates of seond point were in the stereotactic space. Angular accuracy was estimated at $+/-$ 5 degrees when second coordinates were taken from scoutviews, $+/-$ degrees if coordinates were taken from another scanned axial view.

Discussion

The present study describes the development of a methology and computer program to transpose image-defined trajectories into angular settings to be used with arc-quadrant stereotactic systems[1-4]. Frames need to be adapted isocentrically to the CT gantry and parallel to the plane of scanning. A trajectory will be defined by two points coordinates, which can be interactively selected and measured from axial, reformatted or scoutview images. Computer program translate the image generated coordinates into angular coordinates for arc-quadrant centred stereotactic devices. Main advantages of the multiplanar methodology is that scanning can be focused in the region of interest, allowing the performance of high resolution CT in similar time to a conventional head CT scan. At the same time, it allows direct measurement of x, y, z coordinates from any point defined on any axial, reformatted or scoutviews images from CT and on any direct multiplanar MRI image. Target and trajectory selection can be interactively selected, measured and optimized. Display of trajectories on orthogonal antero-posterior and lateral scoutviews constitute another advantage, allowing the intraoperative assessment of trajectories by correlation with intraoperative orthogonal X-rays, ventriculography and angiography. The use of the computer program allows the accurate and reliable transposition of 2 D image-defined trajectories into arc-quadrant centred stereotactic devices. The use of this results in optimization of multiplanar volumetric-oriented stereotaxis, *i.e.* interstitial radiotherapy, laser resection, stereotactic guided resection, etc. and in decrease of complications allowing the selection of safest trajectory to any intracranial lesion.

References

1. Birg W, Mundinger F (1973) Computer calculations of target parameters for a stereotactic apparatus. Acta Neurochir (Wien) 29: 123–129
2. Brown R (1979) A computerized tomography-computergraphics approach to stereotactic localization. J Neurosurg 50: 715–720
3. Goerss S, Kelly P, Kall B *et al* (1982) A computed tomographic stereotactic adaptation system. Neurosurg 10: 375–379
4. Zamorano L, Dujovny M, Malik G (1987) Factors affecting measurements in CT image guided stereotactic procedures. Appl Neurophysiol 50: 53–56

Correspondence: Lucia J. Zamorano, M.D., Ph.D., Henry Ford Neurosurgical Institute, Department of Neurological Surgery, Henry Ford Hospital, 2799 West Grand Boulevard, Detroit, MI 48202, U.S.A.

Acta Neurochirurgica, Suppl. 46, 112–114 (1989)
© by Springer-Verlag 1989

The Stereotactic Operating Microscope: Accuracy Refinement and Clinical Experience*

D. W. Roberts, J. W. Strohbehn, E. M. Friets, J. Kettenberger, and **A. Hartov**

Dartmouth-Hitchcock Medical Center and Thayer School of Engineering, Dartmouth College, Hanover, NH, U.S.A.

Summary

Accuracy of a stereotactic operating microscope, by which imaging data may be superimposed on the operative field without a stereotactic frame, has been most limited by the resolution of imaging information. Using newer algorithms and pilot pole calibration of the digitizer, an error in registration of 2 mm and in contour display of 3 mm has been demonstrated. Greatest utility of the system clinically has been in providing navigational guidance to small lesions undergoing resection.

Keywords: Stereotaxy; operating microscope; CT scanning; MRI; computer.

A computer-based system in which three-dimensional imaging information (CT and MRI) is spatially registered with the operating microscope has been developed, and descriptions of early system design and a prototype have been previously reported[1, 2]. This system integrates radiologic information – such as the location or boundary of a tumour – with the optics of the operating microscope such that the surgeon sees that information superimposed on the surgical field in correct position, orientation, and scale. Using non-imaging ultrasonic range-finders, the system requires neither stereotactic frame nor mechanical linkage between surgical field and operating microscope. Further work has been directed towards improvement in accuracy, and subsequent modifications, phantom testing, and clinical experience are presented.

Accuracy Improvement

Error analysis of the system has enabled appreciation of those steps in its operation at which greatest error is introduced or magnified. The single greatest source of error was found in the resolution of the CT imaging data, which in turn was most limited by the thickness of individual slices. The system (as with any stereotactic system) cannot be more accurate than the

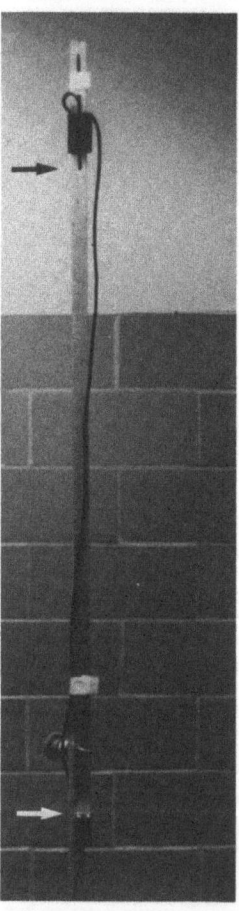

Fig. 1. Pilot pole with spark gap (black arrow) and microphone (white arrow) separated by a known distance

* Supported in part by: The Whitaker Foundation, Camp Hill, PA, and DHHS/NIH grant GM 37308.

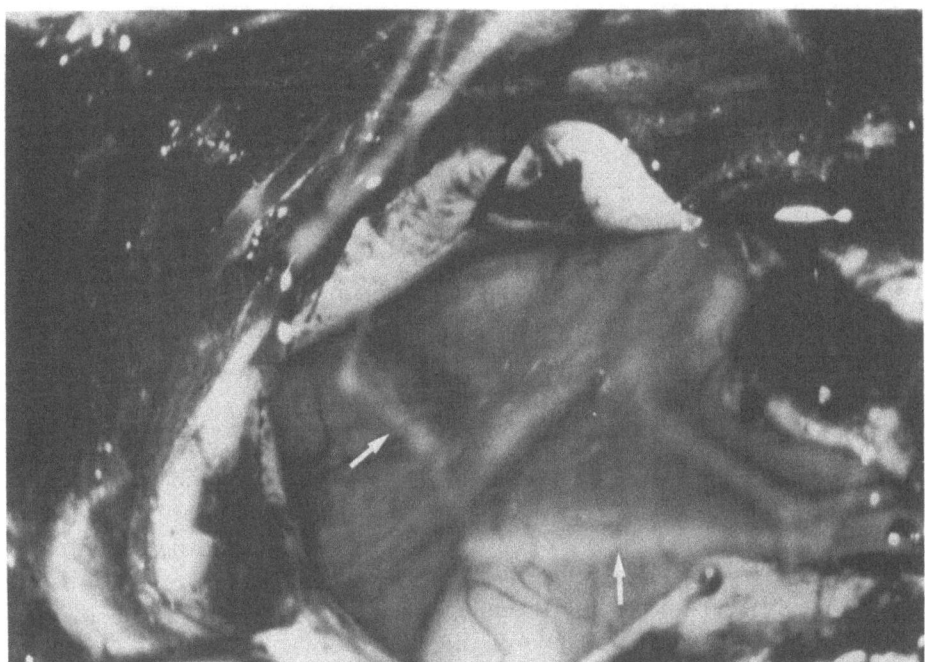

Fig. 2. Intraoperative photograph of the contour (white arrows) of a subcortical mass lesion projected onto the overlying cortical surface, as seen through the operating microscope

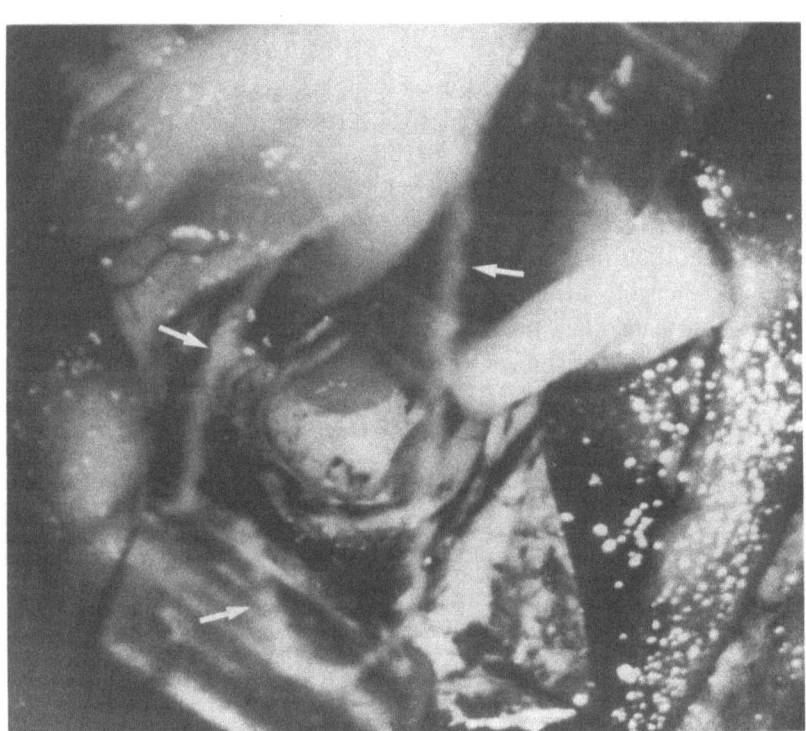

Fig. 3. Intraoperative photograph of the contour (white arrows) of a tumour in the focal plane of the operating microscope

imaging information with which is works. This has encouraged the use of imaging with minimal slice thickness.

A second major source of error was demonstrated to arise from the slant-range measurements of the ultrasonic digitizer. These distances are calculated from the temperature-dependent time of flight of an acoustic impulse from a spark gap (held at a CT-imaged registration point or attached to the microscope) to an overhead microphone. The commerically-configured digitizer measured temperature at one microphone for this calculation, and this accounted poorly for the temperature gradients which characterized the operating room environment. To improve upon this, a vertically-oriented pilot pole of known length was introduced (see Fig. 1). Each time the digitizer now calculates a slant-range, the time of flight between a spark gap and a microphone on the pilot pole is first measured, and that sound velocity is then entered into the subsequent calculations.

In addition to recognizing these errors in the information with which the system must operate, error analysis identified a major source of error magnification in the algorithm for calculation of the focal point. The original system used an operating room coordinate space based upon oblique spark gap coordinates (*e.g.*, the position of each registration point was determined in relation to the microscope's spark gaps). Converting to an operating room coordinate space that is based upon the microphones (whose separation is approximately three times that of the spark gaps) decreased the average error by 34% and the maximum error by 41%.

Phantom Testing

System accuracy has been assessed using two phantoms. The first phantom is a plexiglass block upon which multiple points have been inscribed using a milling machine accurate to 0.025 mm. Using that information rather than imaging data to locate three registration fiducials (and thereby eliminating the contribution of imaging to error), the system's ability to locate a fourth unknown point has had an average error of 1.2 ± 0.4 mm. Projection of a simulated contour has shown an average error of 3.0 ± 0.6 mm.

The second phantom is a plexiglass staircase that has been CT scanned with 1.5 mm slice thickness. Simulating the entire operation of the system as it would be employed clinically, testing has shown an average error of 2.0 ± 0.5 mm in its ability to locate an imaged point. Projection of a contour has had an error of 1.7 ± 1.0 mm.

Clinical Experience

The stereotactic operating microscope has been increasingly unobstrusive and reliable during clinical employment. Mean accuracy in the less controlled environment of the operating room has been within 2–3 mm, and although slightly less than that of phantom testing, this has still enabled the system to be of practical utility in guidance to small lesions. Contours of a subcortical mass projected up onto the cortical surface and of a partially resected tumour in the focal plane of the microscope are illustrated in Figs. 2 and 3 respectively.

References

1. Hatch JF, Roberts DW, Strohbehn JW (1985) Reference-display system for the integration of CT scanning and the operating microscope. In: Kuklinski WS, Ohley WJ (eds) Proceedings of the Eleventh Annual Northeast Bioengineering Conference. IEEE, New York, pp 252–254
2. Roberts DW, Strohbehn JW, Hatch JF, Murray W, Kettenberger H (1986) A frameless stereotaxic integration of computerized tomographic imaging and the operating microscope. J Neurosurg 65: 545–549

Correspondence: David W. Roberts, M.D., Section of Neurosurgery, Dartmouth-Hitchcock Medical Center, Hanover, NH 03756, U.S.A.

Springer-Verlag Wien NewYork

Acta Neurochirurgica

Supplementum 44

1988. 35 partly coloured figures.
VIII, 185 pages.
Cloth DM 220,–, öS 1540,–
Reduced price for subscribers to
"Acta Neurochirurgica":
Cloth DM 198,–, öS 1386,–
ISBN 3-211-82088-4

Supplementum 45

1988. 16 figures. VII, 55 pages.
Cloth DM 56,–, öS 390,–
Reduced price for subscribers to
"Acta Neurochirurgica":
Cloth DM 50,40, öS 351,–
ISBN 3-211-82096-5

J. Brihaye, L. Calliauw, F. Loew,
R. van den Bergh (Eds.)

Personality and Neurosurgery

Proceedings of the Third Convention
of the Academia Eurasiana Neurochirurgica, Brussels,
August 30 – September 2, 1987

The human personality is inextricably
bound up with the function of the central nervous system. Diseases and malfunc-
tions of the brain, head injuries and neurosurgical operations can all result in permanently altered behaviour patterns. This
interrelation between brain and behaviour
is most clearly demonstrated in cases
involving functional neurosurgery and
severe traumatic lesions.
Despite the fact that this interrelation
represents an everyday challenge to the
neurosurgeon, it is a question which re-
ceives less attention than it deserves in
neurosurgical meetings.
At the Third Convention of the Academia
Eurasiana Neurochirurgica the following
topics were discussed:
– Meaning of human personality
– Methodology of personality evaluation
– Changes of personality as consequence
  of severe brain injuries
– Epilepsy and personality
– Aphasia and personality
– Psychosurgery and personality

H.-J. Reulen, J. Philippon (Eds.)

Prevention and Treatment of Delayed Ischaemic Dysfunction in Patients with Subarachnoid Haemorrhage

An Update

The papers presented in this supple-
ment volume of Acta Neurochirurgica
cover a broad spectrum from basic data on
the pathophysiology of Subarachnoid
Haemorrhage (SAH) and delayed ischaemic dysfunction to the clinical use of
Nimodipine which has been largely documented among calcium inhibitors for its
cerebro-vascular properties.
The first two presentations deal with the
pathophysiological events following SAH
and discuss a variety of mechanisms which
may be responsible for the neurological
dysfunction.
The second part discusses the cerebro-
vascular and direct cerebral effects of
Nimodipine and its evaluation by trans-
cranial Doppler sonography and the inter-
actions of this drug with general anaesthesia. Therapeutic aspects constitute the
third part of this volume. On the basis of
present knowledge and experience it seems
that early surgery combined with prophy-
lactic treatment represents the best
option for patients after
aneurysmal rupture.

Acta Neurochirurgica /
Supplementum 44

Acta Neurochirurgica /
Supplementum 45

Moelkerbastei 5, A-1011 Wien · 175 Fifth Avenue, New York, NY 10010, USA ·
Heidelberger Platz 3, D-1000 Berlin · 37-3, Hongo 3-chome, Bunkyo-ku, Tokyo 113, Japan